Springer Undergraduate Mathematics Series

Advisory Board

M.A.J. Chaplain *University of Dundee*
K. Erdmann *Oxford University*
A. MacIntyre *Queen Mary, University of London*
L.C.G. Rogers *University of Cambridge*
E. Süli *Oxford University*
J.F. Toland *University of Bath*

Other books in this series

A First Course in Discrete Mathematics *I. Anderson*
Analytic Methods for Partial Differential Equations *G. Evans, J. Blackledge, P. Yardley*
Applied Geometry for Computer Graphics and CAD, Second Edition *D. Marsh*
Basic Linear Algebra, Second Edition *T.S. Blyth and E.F. Robertson*
Basic Stochastic Processes *Z. Brzeźniak and T. Zastawniak*
Calculus of One Variable *K.E. Hirst*
Complex Analysis *J.M. Howie*
Elementary Differential Geometry *A. Pressley*
Elementary Number Theory *G.A. Jones and J.M. Jones*
Elements of Abstract Analysis *M. Ó Searcóid*
Elements of Logic via Numbers and Sets *D.L. Johnson*
Essential Mathematical Biology *N.F. Britton*
Essential Topology *M.D. Crossley*
Fields and Galois Theory *J.M. Howie*
Fields, Flows and Waves: An Introduction to Continuum Models *D.F. Parker*
Further Linear Algebra *T.S. Blyth and E.F. Robertson*
General Relativity *N.M.J. Woodhouse*
Geometry *R. Fenn*
Groups, Rings and Fields *D.A.R. Wallace*
Hyperbolic Geometry, Second Edition *J.W. Anderson*
Information and Coding Theory *G.A. Jones and J.M. Jones*
Introduction to Laplace Transforms and Fourier Series *P.P.G. Dyke*
Introduction to Lie Algebras *K. Erdmann and M.J. Wildon*
Introduction to Ring Theory *P.M. Cohn*
Introductory Mathematics: Algebra and Analysis *G. Smith*
Linear Functional Analysis *B.P. Rynne and M.A. Youngson*
Mathematics for Finance: An Introduction to Financial Engineering *M. Capiński and T. Zastawniak*
Matrix Groups: An Introduction to Lie Group Theory *A. Baker*
Measure, Integral and Probability, Second Edition *M. Capiński and E. Kopp*
Metric Spaces *M. Ó Searcóid*
Multivariate Calculus and Geometry, Second Edition *S. Dineen*
Numerical Methods for Partial Differential Equations *G. Evans, J. Blackledge, P.Yardley*
Probability Models *J. Haigh*
Real Analysis *J.M. Howie*
Sets, Logic and Categories *P. Cameron*
Special Relativity *N.M.J. Woodhouse*
Symmetries *D.L. Johnson*
Topics in Group Theory *G. Smith and O. Tabachnikova*
Vector Calculus *P.C. Matthews*

James N. Webb

Game Theory

Decisions, Interaction and Evolution

Springer

Library
Quest University Canada
3200 University Boulevard
Squamish, BC V8B 0N8

James N. Webb, BSc, PhD, CPhys, MInstP

Cover illustration elements reproduced by kind permission of:
Aptech Systems, Inc., Publishers of the GAUSS Mathematical and Statistical System, 23804 S.E. Kent-Kangley Road, Maple Valley, WA 98038, USA. Tel: (206) 432 - 7855 Fax (206) 432 - 7832 email: info@aptech.com URL: www.aptech.com.
American Statistical Association: Chance Vol 8 No 1, 1995 article by KS and KW Heiner 'Tree Rings of the Northern Shawangunks' page 32 fig 2.
Springer-Verlag: Mathematica in Education and Research Vol 4 Issue 3 1995 article by Roman E Maeder, Beatrice Amrhein and Oliver Gloor 'Illustrated Mathematics: Visualization of Mathematical Objects' page 9 fig 11, originally published as a CD ROM 'Illustrated Mathematics' by TELOS: ISBN 0-387-14222-3, German edition by Birkhauser: ISBN 3-7643-5100-4.
Mathematica in Education and Research Vol 4 Issue 3 1995 article by Richard J Gaylord and Kazume Nishidate 'Traffic Engineering with Cellular Automata' page 35 fig 2. Mathematica in Education and Research Vol 5 Issue 2 1996 article by Michael Trott 'The Implicitization of a Trefoil Knot' page 14.
Mathematica in Education and Research Vol 5 Issue 2 1996 article by Lee de Cola 'Coins, Trees, Bars and Bells: Simulation of the Binomial Process' page 19 fig 3. Mathematica in Education and Research Vol 5 Issue 2 1996 article by Richard Gaylord and Kazume Nishidate 'Contagious Spreading' page 33 fig 1.
Mathematica in Education and Research Vol 5 Issue 2 1996 article by Joe Buhler and Stan Wagon 'Secrets of the Madelung Constant' page 50 fig 1.

Mathematics Subject Classification (2000): 90C39; 90C40; 91A05; 91A06; 91A10; 91A13; 91A15; 91A18; 91A20; 91A22; 91A25; 91A30; 91A35; 91A40

British Library Cataloguing in Publication Data
A catalogue record for this book is available from the British Library

Library of Congress Control Number: 2006931002

Springer Undergraduate Mathematics Series ISSN 1615-2085
ISBN-10: 1-84628-423-6   e-ISBN-10: 1-84628-636-0   Printed on acid-free paper
ISBN-13: 978-1-84628-423-6   e-ISBN-13: 978-1-84628-636-0

© Springer-Verlag London Limited 2007
The right of James N. Webb to be identified as the author of this work has been asserted in accordance with Sections 77 and 78 of the Copyright Designs and Patents Act 1988.

Apart from any fair dealing for the purposes of research or private study, or criticism or review, as permitted under the Copyright, Designs and Patents Act 1988, this publication may only be reproduced, stored or transmitted, in any form or by any means, with the prior permission in writing of the publishers, or in the case of reprographic reproduction in accordance with the terms of licences issued by the Copyright Licensing Agency. Enquiries concerning reproduction outside those terms should be sent to the publishers.

The use of registered names, trademarks, etc. in this publication does not imply, even in the absence of a specific statement, that such names are exempt from the relevant laws and regulations and therefore free for general use.

The publisher makes no representation, express or implied, with regard to the accuracy of the information contained in this book and cannot accept any legal responsibility or liability for any errors or omissions that may be made.

9 8 7 6 5 4 3 2 1

Springer Science+Business Media
springer.com

Preface

This book is an introduction to game theory from a mathematical perspective. It is intended to be a first course for undergraduate students of mathematics, but I also hope that it will contain something of interest to advanced students or researchers in biology and economics who often encounter the basics of game theory informally via relevant applications. In view of the intended audience, the examples used in this book are generally abstract problems so that the reader is not forced to learn a great deal of a subject – either biology or economics – that may be unfamiliar. Where a context is given, these are usually "classical" problems of the subject area and are, I hope, easy enough to follow.

The prerequisites are generally modest. Apart from a familiarity with (or a willingness to learn) the concepts of a proof and some mathematical notation, the main requirement is an elementary understanding of probability. A familiarity with basic calculus would be useful for Chapter 6 and some parts of Chapters 1 and 8. The basic ideas of simple ordinary differential equations are required in Chapter 9 and, towards the end of that chapter, some familiarity with matrices would be an advantage – although the relevant ideas are briefly described in an appendix.

I have tried to provide a unified account of single-person decision problems ("games against nature") as well as both classical and evolutionary game theory, whereas most textbooks cover only one of these. There are two immediate consequences of this broad approach. First, many interesting topics are left out. However, I hope that this book will provide a good foundation for further study and that the books suggested for further reading at the end of this volume will go some way to filling the gaps. Second, the notation and terminology used may be different in places from that which is commonly used in each of the three separate areas. In this book, I have tried to use similar (combinations of) symbols to represent similar concepts in each part, and it should be clear from the context what is meant in any particular case.

If time is limited, lecturers could make selections of the material according to the interests and mathematical background of the students. For example, a course on non-evolutionary game theory could include material from Chapters 1, 2, and 4–7. A course on evolutionary game theory could include material from Chapters 1, 2, 4, 8, and 9.

Finally, it is a pleasure to thank Vassili Kolokoltsov, Hristo Nikolov, and two anonymous reviewers whose perceptive comments have helped to improve this book immeasurably. Any flaws that remain are, of course, the responsibility of the author alone.

Nottingham   James Webb
May 2006

Contents

Part I. Decisions
---

1. **Simple Decision Models** 3
   1.1 Optimisation... 3
   1.2 Making Decisions 5
   1.3 Modelling Rational Behaviour 11
   1.4 Modelling Natural Selection 17
   1.5 Optimal Behaviour 21

2. **Simple Decision Processes** 23
   2.1 Decision Trees 23
   2.2 Strategic Behaviour 24
   2.3 Randomising Strategies 27
   2.4 Optimal Strategies 31

3. **Markov Decision Processes** 37
   3.1 State-dependent Decision Processes 37
   3.2 Markov Decision Processes 39
   3.3 Stochastic Markov Decision Processes 42
   3.4 Optimal Strategies for Finite Processes 46
   3.5 Infinite-horizon Markov Decision Processes 48
   3.6 Optimal Strategies for Infinite Processes 50
   3.7 Policy Improvement 54

Part II. Interaction
---

4. Static Games 61
4.1 Interactive Decision Problems 61
4.2 Describing Static Games 63
4.3 Solving Games Using Dominance 66
4.4 Nash Equilibria 68
4.5 Existence of Nash Equilibria 76
4.6 The Problem of Multiple Equilibria 78
4.7 Classification of Games 80
4.8 Games with n-players 86

5. Finite Dynamic Games 89
5.1 Game Trees 89
5.2 Nash Equilibria 91
5.3 Information Sets 93
5.4 Behavioural Strategies 95
5.5 Subgame Perfection 99
5.6 Nash Equilibrium Refinements 101

6. Games with Continuous Strategy Sets 107
6.1 Infinite Strategy Sets 107
6.2 The Cournot Duopoly Model 107
6.3 The Stackelberg Duopoly Model 111
6.4 War of Attrition 114

7. Infinite Dynamic Games 119
7.1 Repeated Games 119
7.2 The Iterated Prisoners' Dilemma 121
7.3 Subgame Perfection 125
7.4 Folk Theorems 129
7.5 Stochastic Games 132

Part III. Evolution

8. Population Games 139
8.1 Evolutionary Game Theory 139
8.2 Evolutionarily Stable Strategies 140
8.3 Games Against the Field 144
8.4 Pairwise Contest Games 148
8.5 ESSs and Nash Equilibria 153
8.6 Asymmetric Pairwise Contests 157
8.7 Existence of ESSs 160

9. Replicator Dynamics 165
- 9.1 Evolutionary Dynamics 165
- 9.2 Two-strategy Pairwise Contests 168
- 9.3 Linearisation and Asymptotic Stability 171
- 9.4 Games with More Than Two Strategies 174
- 9.5 Equilibria and Stability 179

Part IV. Appendixes

A. Constrained Optimisation 189

B. Dynamical Systems 193

Solutions 205

Further Reading 235

Bibliography 237

Index 239

Part I

Decisions

1
Simple Decision Models

1.1 Optimisation

Suppose we are faced with the problem of making a decision. One approach to the problem might be to determine the desired outcome and then to behave in a way that leads to that result. This leaves open the question of whether it is always possible to achieve the desired outcome. An alternative approach is to list the courses of action that are available and to determine the outcome of each of those behaviours. One of these outcomes is preferred because it is the one that maximises[1] something of value (for example, the amount of money received). The course of action that leads to the preferred outcome is then picked from the available set. We will call the second approach "making an optimal decision". In this book, we will develop a mathematical framework for studying the problem of making an optimal decision in a variety of circumstances.

Finding the maximum of something is a familiar procedure in basic calculus. Suppose we are interested in finding the maximum of some function, let's call it $f(x)$. We differentiate f and set the result equal to zero. A solution of this equation gives us one or more values of x at which a maximum is attained, which we might call x^*. The maximum value of the function is then $f(x^*)$. (We must also check that the value of the second derivative of f to make sure that $f(x^*)$ is really a maximum.)

In basic calculus, it is usually assumed (often without being mentioned) that

---

[1] In some situations, the aim may be to minimise a loss. In this case, we can still talk about maximisation by considering the negative of the loss as the outcome.

the function we wish to maximise is defined for all real values of x (i.e., $x \in \mathbb{R}$) and is continuous. However, we will encounter problems in which one or both of these assumptions are untrue: the function $f(x)$ may only be defined for values of x in a compact set $\mathbf{X}$; the set $\mathbf{X}$ may be discrete; or the function being maximised may be discontinuous by definition. These distinctions can have important consequences, as is shown by the following example and exercises.

Example 1.1

Consider the function $f(x) = \sqrt{x}$. If this function is defined for all $x \in [0, \infty)$, then it keeps increasing as x increases and so it does not have a maximum. However, if the function is only defined for $x \in [0, 4]$, then the function does have a maximum. The maximum value of f is attained at one boundary $x^* = 4$ and $f(x^*) = 2$.

Exercise 1.1

Maximise $F(n) = 1 - n^2 + \frac{9}{2}n$ where n is integer. (Remember that n^* must be an integer.)

Exercise 1.2

Let a, b and c be positive constants. Let

$$f(x) = \begin{cases} \left(1 - \dfrac{x}{b}\right) & \text{if } 0 \leq x \leq b \\ 0 & \text{otherwise} \end{cases}$$

and

$$\chi(x) = \begin{cases} 1 & \text{if } x > 0 \\ 0 & \text{if } x \leq 0 \end{cases}.$$

Maximise $g(x) = axf(x) - c\chi(x)$.

We will often wish to focus on the value of x at which the maximum is achieved rather than the maximum value of the function itself, so we introduce a new symbol *argmax*.

Definition 1.2

Suppose x is an arbitrary member of some set $\mathbf{X}$. Let $f(x)$ be some function that is defined $\forall x \in \mathbf{X}$. Then the symbol *argmax* is defined by the following equivalence.

$$x^* \in \operatorname*{argmax}_{x \in \mathbf{X}} f(x) \quad \Longleftrightarrow \quad f(x^*) = \max_{x \in \mathbf{X}} f(x).$$

That is, x^* is a value that maximises the function $f(x)$. Note that we do not write $x^* = \text{argmax}_{x \in \mathbf{X}} f(x)$ because a function may take its maximum value for more than one element in the set $\mathbf{X}$. Because the symbol *argmax* returns a set of values rather than a unique value, it is called a correspondence rather than a function.

Example 1.3

Consider the function defined on $x \in [-2, 2]$ by $f(x) = x^2$. This function achieves its maximum at $x^* = \pm 2$. So

$$\underset{x \in [-2,2]}{\text{argmax}} \, x^2 = \{+2, -2\} \,.$$

Exercise 1.3

(a) Let $f(x) = 1 + 6x - x^2$ be defined $\forall x \in \mathbb{R}$. Find $\text{argmax}_{x \in \mathbb{R}} f(x)$.
(b) Let $f(x) = 1 + 6x - x^2$ be defined $\forall x \in [1, 2]$. Find $\text{argmax}_{x \in [1,2]} f(x)$.
(c) Let $f(x) = (1 - x)^2$ be defined $\forall x \in [0, 3]$. Find $\text{argmax}_{x \in [0,3]} f(x)$.

1.2 Making Decisions

The simplest case to consider is when there is no randomness in the environment – once a choice has been made, the outcome is certain. To begin to build a theory of optimal decisions, we make the following definitions.

Definition 1.4

A choice of behaviour in a single-decision problem is called an *action*. The set of alternative actions available will be denoted $\mathbf{A}$. This will either be a discrete set, e.g., $\{a_1, a_2, a_3, \ldots\}$, or a continuous set, e.g., the unit interval $[0, 1]$.

Definition 1.5

A *payoff* is a function $\pi \colon \mathbf{A} \to \mathbb{R}$ that associates a numerical value with every action $a \in \mathbf{A}$.

Definition 1.6

An action a^* is an *optimal action* if

$$\pi(a^*) \geq \pi(a) \quad \forall a \in A \qquad (1.1)$$

or, equivalently,

$$a^* \in \underset{a \in A}{\operatorname{argmax}}\, \pi(a)\,. \qquad (1.2)$$

That is, the optimal decision is to choose an $a^* \in \mathbf{A}$ that maximises the payoff $\pi(a)$. In general, a^* need not be a unique choice of action: if two actions lead to the same, maximal payoff, then either will do (notice the weak inequality in the first form of the definition).

Example 1.7

A jobseeker is offered two jobs, J_1 and J_2. Their possible actions are a_i = accept J_i with $i = 1, 2$. The payoffs are the salaries on offer: J_1 pays £15000, J_2 pays £17000. Because $\pi(a_1) = 15000$ and $\pi(a_2) = 17000$ the optimal decision is $a^* = a_2$ (i.e., accept the second job).

Exercise 1.4

An investor going to invest £1000 for a year and has narrowed the choice to one of two savings accounts. The two accounts differ only in the rate of return: the first pays 6% annually, and the second pays 3% at six month intervals. Which account should the investor choose? Does the answer depend on whether or not the initial capital is included in the payoff?

In the previous example and exercise, the optimal decisions are not altered if payoffs are given in U.S. dollars (or any other currency) rather than pounds; nor are they altered if £1000 is added to each payoff. These alterations to the payoffs are both examples of affine transformations.

Definition 1.8

An *affine transformation* changes payoffs $\pi(a)$ into payoffs $\pi'(a)$ according to the rule

$$\pi'(a) = \alpha \pi(a) + \beta$$

where α and β are constants independent of a and $\alpha > 0$.

Theorem 1.9

The optimal action is unchanged if payoffs are altered by an affine transformation.

Proof

Because $\alpha > 0$ we have

$$\operatorname*{argmax}_{a \in \mathbf{A}} \pi'(a) = \operatorname*{argmax}_{a \in \mathbf{A}} [\alpha \pi(a) + \beta]$$
$$= \operatorname*{argmax}_{a \in \mathbf{A}} \pi(a) \ .$$

$\square$

We now consider some problems in which the action set is a continuous subset of $\mathbb{R}$. This can arise as a convenient approximation for models in which a discrete action set has a large number of elements. For example, if we are selling something, we might want to consider charging prices between £0.01 and £5.00. Because we can only charge prices in whole pennies, the action set is discrete. But, rather than consider the consequences of 500 separate actions, we treat price as continuous and employ the powerful features of calculus to solve the problem.

Example 1.10

The Convent Fields Soup Company makes tomato soup. If it charges a price of p pounds per litre, then the market will buy $Q(p)$ litres, where

$$Q(p) = \begin{cases} Q_0 \left(1 - \frac{p}{p_0}\right) & \text{if } p < p_0 \\ 0 & \text{if } p \geq p_0 \end{cases}.$$

$Q(p)$ is a non-increasing function of price p. So Q_0 is a constant that gives the maximum quantity that could be sold (at any price), and p_0 is a constant that gives the maximum price that the market would be prepared to pay. The actions available to the company are the choice of a price $p \in [0, p_0]$. There is no point in setting a price above p_0 because the company would sell no soup. Suppose that the cost of producing soup is c pounds per litre. Taking the company's profit as its payoff, we have $\pi(p) = (p - c)Q(p)$. The optimal decision is to set a price p^* that maximises the profit. To find this price, we find the maximum of the payoff as a function of price. Because

$$\frac{d\pi}{dp}(p^*) = Q_0 \left(1 + \frac{c}{p_0} - \frac{2p^*}{p_0}\right) = 0$$

the optimal action is, therefore, to choose a price $p^* = \frac{1}{2}(p_0 + c)$.

Example 1.11

In the previous example, we assumed that the soup factory has already been constructed. The problem changes if the decision is being made before the factory is built. For simplicity, let us assume that the marginal cost of production is zero (the more general case is covered in Exercise 1.6) and suppose that it costs an amount B to build the factory. The payoff then appears to be $\pi(p) = pQ(p) - B$. Because B is a constant, the optimal price is $p^* = \frac{1}{2}p_0$. However, if soup is sold at this price, the profit made by the company is

$$\pi(p^*) = \frac{Q_0 p_0}{4} - B$$

If B is large enough the company could make a loss by selling soup at this price. The optimal action is, therefore, to choose a price

$$p^* = \begin{cases} \frac{1}{2}(p_0 + c) & \text{if the profit will be positive} \\ 0 & \text{otherwise.} \end{cases}$$

Exercise 1.5

A company makes small widgets. If the manufacturer produces q widgets per day, they can be sold at a price $P(q)$, where

$$P(q) = P_0 \max\left\{\left(1 - \frac{q}{q_0}\right), 0\right\}.$$

Assume the number of widgets produced is very large, so q can be treated as a continuous variable. (a) What quantity should be made to maximise the manufacturer's income? (b) If manufacturing costs increase linearly with the number of widgets made (i.e., cost $= cq$), what quantity maximises the manufacturer's profit?

Exercise 1.6

A company is considering building a factory to make fertilizer. At a price p, $T(p)$ tonnes will be sold, where

$$T(p) = T_0 \max\left\{\left(1 - \left(\frac{p}{p_0}\right)\right), 0\right\}.$$

Suppose manufacturing costs increase with the tonnage made, t, as $C(t) = c_0 + c_1 t$ where c_0 and c_1 are non-negative constants. What price would maximise the manufacturer's income? Should the company build the factory?

1.2 Making Decisions

So far, we have assumed that there is no uncertainty about the consequences of any decisions. If uncertainty exists, we can compare the expected outcome (in the probabilistic sense) for each action. Uncertainty about payoffs can be represented as a random variable, X, which takes certain values corresponding to possible "states of Nature" (e.g., economic conditions) with specified probabilities. We will denote the set of "states of Nature" by $\mathbf{X}$ and the probability with which a particular state x occurs will be denoted $P(X = x)$. If the payoff associated with action a when the state of Nature is x is $\pi(a|x)$, then the payoff for adopting action a is

$$\pi(a) = \sum_{x \in \mathbf{X}} \pi(a|x) P(X = x)$$

and an optimal action is

$$a^* \in \operatorname*{argmax}_{a \in A} \sum_{x \in \mathbf{X}} \pi(a|x) P(X = x) \,.$$

Example 1.12

An investor has £1000 to invest for one year. Their[2] available actions are

a_1: Put the money in a building society account that yields 7% interest p.a.

a_2: Invest in a share fund that gives a return of £1500 if the stock market performs well and £600 (i.e., a loss of £400) if the stock market performs badly.

The state of Nature is the performance of the stock market, which is good 50% of the time and bad for the remaining 50% of the time. So we have the set of states $\mathbf{X} = \{Good, Bad\}$ with $P(X = Good) = P(X = Bad) = 0.5$. The expected payoffs (in pounds) for the two possible actions are

$$\begin{aligned} \pi(a_1) &= 1070 \\ \pi(a_2) &= \frac{1}{2} 1500 + \frac{1}{2} 600 = 1050. \end{aligned}$$

So, the optimal action is a_1 (put the money in the building society).

---

[2] As always, there is a problem in writing about individuals whose gender is irrelevant: which pronoun to use? Rather than make an invidious choice or use the cumbersome "he or she", I have opted to use "they", "them", and "their". For example, "they should use the following strategy". The use of these grammatically plural pronouns to refer to an individual is common in colloquial English and the mismatch between grammatical and actual number also occurs in other languages (for example, the polite use of "vous" in French).

Exercise 1.7

Consider the following table of payoffs $\pi(a|x)$ for action set $\mathbf{A} = \{a_1, a_2, a_3\}$ and states of nature $\mathbf{X} = \{x_1, x_2, x_3, x_4\}$.

	x_1	x_2	x_3	x_4
a_1	3	0	3	0
a_2	0	3	0	3
a_3	1	1	1	1

What are the optimal actions if
(a) $P(X = x_1) = P(X = x_2) = P(X = x_3) = P(X = x_4) = \frac{1}{4}$
(b) $P(X = x_1) = P(X = x_3) = \frac{1}{8}$ and $P(X = x_2) = P(X = x_4) = \frac{3}{8}$?

If X is a continuous random variable, then we use a density function $f(x)$ with $P(x < X \leq x + dx) \equiv f(x)dx$. Then the expected payoff for adopting action a is

$$\pi(a) = \int_{x \in \mathbf{X}} \pi(a|x) f(x) \, dx$$

and an optimal action is

$$a^* \in \underset{a \in A}{\operatorname{argmax}} \int_{x \in \mathbf{X}} \pi(a|x) f(x) \, dx \ .$$

Example 1.13

Suppose that an investor has a choice between two investments a_1 and a_2 with payoffs $\pi(a_1|x) = w(1+r)$ and $\pi(a_2|x) = w + X$ where X is a normally distributed random variable, $X \sim N(\mu, \sigma^2)$. For example, a_1 could represent putting an initial capital w into a savings account with interest rate r and a_2 could represent investing the same amount in the stock market. The expected payoffs for the two actions are $\pi(a_1) = w(1+r)$ and

$$\begin{aligned}\pi(a_2) &= w + \int_{-\infty}^{+\infty} x f(x) \, dx \\ &= w + \mu\end{aligned}$$

so the optimal action is

$$a^* = \begin{cases} a_1 & \text{if } wr > \mu \\ a_2 & \text{if } wr < \mu \end{cases}$$

If $wr = \mu$ then the investor is indifferent between a_1 and a_2 (i.e., both a_1 and a_2 are optimal).

Exercise 1.8

Each day a power company produces u units of power at a cost of c dollars per unit and sells them at a price p dollars per unit. Suppose that demand for power is exponentially distributed with mean d units, i.e.,

$$f(x) = \frac{1}{d}\exp\left(-\frac{x}{d}\right).$$

If demand exceeds supply, then the company makes up the shortfall by buying units from another company at a cost of k dollars per unit ($k > c$). Show that the expected profit for the company (in dollars) is

$$\pi(u) = pd - cu - kde^{-u/d}$$

and find the optimal level of production.

1.3 Modelling Rational Behaviour

Suppose a person is approached by a wealthy philanthropist who offers them a choice between getting £1 for certain and a 50% chance of getting £3 (and a 50% chance of nothing). Should the person choose the certain outcome or the gamble? What should they choose if the sums involved were £1 million and £3 million? Based on our procedure from the previous section, we might be tempted to say that the person "should" choose the gamble in each case because the expected amount of money received is higher than the amount of money received if the gamble is refused. However, most people will gamble with the low amounts but go for the certain million. Are they being inconsistent or irrational?

Is maximising expected monetary value (EMV) what people *should* do? There are several reasons why they might not. First, people value things other than money: holidays, health, happiness, even the well-being of other people. Second, for most people money is only a means to an end so the "real" value of an amount of money need not be equal to its face value. Consider the case of the wealthy philanthropist who is offering the choice involving millions of pounds. Receiving £1 million will allow the recipient to retire and not have to worry about pensions or life insurance. Receiving £3 million is better than receiving £1 million but it isn't 3 times as good because it is not possible to retire three times over. Third, the reaction to uncertainty may depend on personal circumstances.

Example 1.14

When it sells some insurance, a company assesses the probabilities of various payouts. From this it calculates its expected loss L. To this loss it adds its profit P and charges $C = L + P$. Assuming that a customer agrees with the company about the expected loss L, why do they buy insurance (given that $C > L$)? Part of the reason may be that, although they can afford the expected loss, they could not afford the actual loss if it occurred. So, maximising EMV is appropriate for one "individual" (the insurance company), but not for another (the customer).

Apart from these considerations, there is another – more important – reason why we should not define rationality as maximising EMV: it switches the origin of a behaviour with its consequence. What we would like to do is define rationality in some way and then determine, as a consequence of this definition, whether we can derive *any* quantity that rational people will maximise. If we can do this, then we can use the procedures we have begun to develop with the quantity we have found taking the place of EMV in our calculations. So how can we define rationality?

The first thing to note is that rationality should not be equated with dispassionate reasoning (notwithstanding the view held by certain aliens from a popular science fiction series). An individual's desires lead to a ranking of outcomes in terms of *preference*. These preferences need not accord with those of another individual; however, they should be internally consistent, if they are to form a basis for choice. Thus we will define a rational person as one who has consistent preferences concerning outcomes and will attempt to achieve a preferred outcome.

Suppose we have a set of possible outcomes (for example, eating a hamburger or eating a salad). When asked, people will express preferences concerning these outcomes. These preferences are not necessarily the same for all people: some may prefer the salad while others prefer the hamburger. Given a free choice, people should choose their preferred outcome. (For the moment we will assume that there is no uncertainty about the consequence of a choice: choosing to act in a particular way definitely leads to the desired outcome.) Anyone who really prefers the hamburger but then chooses to eat a salad would be acting "irrationally". Someone who *says* they prefer the hamburger but then chooses to eat a salad because they are on a diet has not expressed their *true* preferences because they have failed to include their desire to lose weight.

Let the set of possible outcomes be denoted by $\Omega = \{\omega_1, \omega_2, \omega_3, \ldots\}$. We will write $\omega_1 \succ \omega_2$ if an individual *strictly* prefers outcome ω_1 over outcome ω_2. We will write $\omega_1 \sim \omega_2$ if an individual is indifferent between the two outcomes. Weak preference will be expressed by the operator $\succeq$. The expression $\omega_1 \succeq \omega_2$

1.3 Modelling Rational Behaviour

means that an individual either prefers ω_1 to ω_2 or is indifferent between the two outcomes.

Definition 1.15

An individual will be called *rational under certainty* if their preferences for outcomes satisfy the following conditions:

1. (Completeness) Either $\omega_1 \succeq \omega_2$ or $\omega_2 \succeq \omega_1$.
2. (Transitivity) If $\omega_1 \succeq \omega_2$ and $\omega_2 \succeq \omega_3$ then $\omega_1 \succeq \omega_3$.

The completeness condition ensures that all outcomes can be compared with each other. The transitivity condition implies that outcomes can be listed in order of preference (possibly with ties between some outcomes). Together these conditions imply that we can introduce the idea of a utility function. An individual will be assumed to have a personal utility function $u(\omega)$ that gives their utility for any outcome, ω. The outcome ω may be numeric (e.g., an amount of money or a number of days of holiday) or less tangible (e.g., degree of happiness). Whatever the reward is, the utility function assigns a number to that reward and encapsulates everything about an outcome that is important to the particular individual being considered.

Definition 1.16

A *utility function* is a function $u: \Omega \to \mathbb{R}$ such that:

$$u(\omega_1) > u(\omega_2) \iff \omega_1 \succ \omega_2$$
$$u(\omega_1) = u(\omega_2) \iff \omega_1 \sim \omega_2$$

An immediate consequence of this definition is that an individual who is rational under certainty should seek to maximise their utility. The relation between the utility function u and the payoff function π is straightforward. Suppose choosing action a produces outcome $\omega(a)$ then $\pi(a) = u(\omega(a))$.

Now let us consider what happens when an action does not produce a definite outcome and instead we allow each outcome to occur with a known probability. Such uncertain outcomes will be called "lotteries".

Definition 1.17

A *simple lottery*, λ, is a set of probabilities for the occurrence of every $\omega \in \Omega$. We shall denote the probability that outcome ω occurs in lottery λ by $p(\omega|\lambda)$.

The set of all possible lotteries will be denoted $\mathbf{\Lambda}$. (Although the set of lotteries depends on the basic set of outcomes $\mathbf{\Omega}$, we will not make this dependence explicit.)

Definition 1.18

A *compound lottery* is a linear combination of simple lotteries (from the same set $\mathbf{\Lambda}$). For example, $q\lambda_1 + (1-q)\lambda_2$ with $0 \leq q \leq 1$ is a compound lottery.

A compound lottery can be regarded as a lottery in which the outcomes are themselves lotteries. The example lottery given in the definition could be taken to mean that simple lottery λ_1 occurs with probability q and simple lottery λ_2 occurs with probability $1-q$. Compound lotteries are not really different from simple lotteries: the compound lottery $q\lambda_1 + (1-q)\lambda_2$ is equivalent to a simple lottery λ with probabilities $p(\omega|\lambda)$, which can be determined from the probabilities $p(\omega|\lambda_1)$, $p(\omega|\lambda_2)$ and the parameter q. However, the ability to define lotteries as combinations of other lotteries is useful for the definition of rationality.

Definition 1.19

An individual will be called *rational under uncertainty* or just *rational* if their preferences for lotteries satisfy the following conditions:

1. (Completeness) Either $\lambda_1 \succeq \lambda_2$ or $\lambda_2 \succeq \lambda_1$.

2. (Transitivity) If $\lambda_1 \succeq \lambda_2$ and $\lambda_2 \succeq \lambda_3$ then $\lambda_1 \succeq \lambda_3$.

3. (Monotonicity) If $\lambda_1 \succ \lambda_2$ and $q_1 > q_2$ then $q_1\lambda_1 + (1-q_1)\lambda_2 \succ q_2\lambda_1 + (1-q_2)\lambda_2$.

4. (Continuity) If $\lambda_1 \succeq \lambda_2$ and $\lambda_2 \succeq \lambda_3$ then there exists a probability q such that $\lambda_2 \sim q\lambda_1 + (1-q)\lambda_3$.

5. (Independence) If $\lambda_1 \succ \lambda_2$ then $q\lambda_1 + (1-q)\lambda_3 \succ q\lambda_2 + (1-q)\lambda_3$

As above, the completeness condition ensures that all lotteries can be compared with each other and the transitivity condition implies that lotteries can be listed in order of preference (possibly with ties). The monotonicity and continuity conditions assert that a lottery gets better smoothly as the probability of a preferred outcome increases. The independence condition implies that preferences only depend on the differences between lotteries; components that are the same can be ignored.

Suppose that choosing action a produces lottery $\lambda(a)$. What is the payoff $\pi(a)$ that a rational individual will seek to maximise? We might try to introduce a utility function for the expected outcome $\mathbb{E}(\omega)$. There are two problems with this. First, it is not clear what $\mathbb{E}(\omega)$ would mean for outcomes that are not given numerically. Second, even when the outcomes are given numerically, it seems that people do not necessarily maximise any function of the expected outcome. Consider the example of wealthy philanthropist again. If an individual takes the gamble for the £3 million, then the expected outcome is £1.5 million. Because people prefer the certain million to the gamble, it would seem – if they are maximising some sort of "utility of the expected outcome" – that the utility of £1 million is greater than the utility of £1.5 million, which seems highly unlikely. An alternative to maximising the "utility" of the expected outcome is maximising the expected utility.

Theorem 1.20 (Expected Utility Theorem)

If an individual is rational in the sense of Definition 1.19, then we can define a utility function $u: \Omega \to \mathbb{R}$ and rational individuals act in a way that maximises the payoff function $\pi(a)$ (the *expected utility*) given by

$$\pi(a) = \sum_{\omega \in \Omega} p(\omega|\lambda(a)) u(\omega) . \tag{1.3}$$

Proof

A more detailed discussion and proof of the Expected Utility Theorem is given by Myerson (1991). □

Remark 1.21

The conditions for rationality expressed in Definition 1.19 only determine the utility function up to an affine transformation (see Definition 1.8). However, this does not present a problem, because the optimality of any behaviour is not altered by a change of this type (see Theorem 1.9).

The explicit construction of a utility function, which is important for constructing realistic models of a person's behaviour (and is a problem that must be solved for each model) will be ignored in this book. We will either assume that maximising EMV is appropriate or specify a utility function. We will also consider completely abstract situations and, in these cases, an individual's payoff will be tacitly specified in "units of utility" without worrying about the

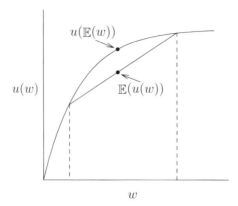

Figure 1.1 The utility function for the individual considered in Exercise 1.9. The utility $u(w)$ of wealth w is such that $\mathbb{E}(u(\omega)) < u(\mathbb{E}(\omega))$, so this individual is risk averse.

various components of an outcome that actually determine this value.

Exercise 1.9

Consider an individual whose utility function of wealth, w, is given by $u(w) = 1 - \exp(-kw)$ with $k > 0$. Assuming that wealth increments are Normally distributed, show that an individual's expected utility can be represented as a trade-off between mean and variance, as in Equation (1.4).

Definition 1.22

An individual whose utility function satisfies $\mathbb{E}(u(\omega)) < u(\mathbb{E}(\omega))$ is said (assuming $\mathbb{E}(\omega)$ can be defined) to be *risk averse*. If $\mathbb{E}(u(\omega)) > u(\mathbb{E}\omega)$ the individual is said to be *risk prone*. If $\mathbb{E}(u(\omega)) = u(\mathbb{E}\omega)$ the individual is said to be *risk neutral*.

Example 1.23

The individual considered in Exercise 1.9 is risk averse. (See Figure 1.1.)

Example 1.24

Consider the following classical portfolio choice problem. Two assets are available to an investor. One is riskless (e.g., a bank account) providing a fixed return of r on the initial sum; the other is risky (e.g., stock market) with a return, having a mean μ and a standard deviation σ. If the investor is a straightforward EMV-maximiser, then they should invest all of their money in stocks if $\mu > r$. However, in some circumstances, a risk-averse investor may prefer to trade-off the expected return and its variance in a linear fashion (see Exercise 1.9). In other words, they can reduce the variability of their return by constructing a portfolio in which they place some fraction of their money in the bank and invest the remainder in the stock market. If a is the fraction that they place in stocks, then the expected return on the portfolio is $a\mu + (1-a)r$ and its variance is $a^2\sigma^2$. So the investor's expected utility is

$$\pi(a) = a\mu + (1-a)r - \frac{k}{2}a^2\sigma^2 \qquad (1.4)$$

where k represents the value that the investor places on the variance relative to the expectation. This expected utility is maximised for

$$a^* = \begin{cases} 0 & \text{if } \mu < r \\ \dfrac{\mu - r}{k\sigma^2} & \text{if } 0 < \mu - r < k\sigma^2 \\ 1 & \mu - r > k\sigma^2 \end{cases}.$$

(Check that the second derivative is negative; or calculate $\pi(a^*)$, $\pi(0)$ and $\pi(1)$.)

Exercise 1.10

Consider an individual whose utility function of wealth, w, is quadratic: $u(w) = w - kw^2$, where the constant k is such that $u(w)$ is non-decreasing over the allowed range for w. Repeat the portfolio problem from Example 1.24.

1.4 Modelling Natural Selection

In this section, we will consider the – at first, rather surprising – proposition that the mathematics describing optimal decisions by rational individuals can also be applied to the behaviour of animals.

The assertion of optimal behaviour by animals rests on the following interpretation of Natural Selection. In the past, a population of animals from a single species contained several types of individual that were genetically programmed to use one of a variety of behaviours.[3] Some of these behaviours result in the animals having few descendants, other behaviours result in animals having many descendants. Through their genes, parents pass on their programmed behaviour to their offspring and after many generations, the type of animal that leaves the greatest number of descendants will be numerically dominant in the population.

Example 1.25

Consider a population consisting of two types of individual, labelled $i = 1, 2$. The animals live for a year, breed once and then die. Individuals of type i have r_i offspring, where we assume (without loss of generality) that $r_1 > r_2$. Suppose that at time t there are $n_i(t)$ animals of type i, then at time $t + 1$ (i.e., the following year) there will be $n_i(t+1) = r_i n_i(t)$ of each type. Starting from time $t = 0$ when there are $n_i(0)$ animals of each type, there are

$$
\begin{aligned}
&n_i(1) = r_i n_i(0) &&\text{at time } t = 1; \\
&n_i(2) = r_i n_i(1) = r_i^2 n_i(0) &&\text{at time } t = 2; \\
&n_i(3) = r_i n_i(2) = r_i^2 n_i(1) = r_i^3 n_i(0) &&\text{at time } t = 3; \\
&\quad\vdots &&\quad\vdots \\
&n_i(t) = r_i n_i(t-1) = \cdots = r_i^t n_i(0) &&\text{at time } t.
\end{aligned}
$$

So the ratio of the numbers of the two types at time t is

$$\frac{n_2(t)}{n_1(t)} = \left(\frac{r_2}{r_1}\right)^t \frac{n_2(0)}{n_1(0)}$$

This ratio tends to zero as $t \to \infty$. In other words, the population comes to be dominated by animals of type 1. We can paraphrase the action of Natural Selection by saying that the animals should "choose" the behaviour that gives the reproduction rate r_1.

Exercise 1.11

Duck-billed platypuses lay n eggs, where n is a characteristic that varies between individuals and is inherited by a platypus's offspring. The probability that each egg hatches is $H(n) = 1 - kn^2$ where $k = 0.1$. After many generations of Natural Selection how many eggs will platypuses be laying? (Remember that n is an integer.)

---

[3] Actually the relationship between genetics and behaviour may be quite complicated and is, in general, poorly understood. The procedure of treating particular behaviours as heritable units is called *the phenotypic gambit* and has proved to be a useful starting point.

1.4 Modelling Natural Selection

Definition 1.26

The *fitness* of a behaviour is defined to be the asymptotic growth rate of the sub-population of animals using that behaviour. That is, for animals with behavioural type i we can define an annual growth rate as

$$r_i(t) = \frac{n_i(t+1)}{n_i(t)}$$

and the fitness for this type is given by $\pi(i) = \lim_{t\to\infty} r_i(t)$.

It is, therefore, a matter of definition that Natural Selection acts in such a way as to maximise fitness. If we want to explain (in evolutionary terms) the behaviour of animals, we consider a set of plausible alternative behaviours and find the one that maximises fitness: this is the behaviour that animals should be using (provided Natural Selection has had enough time to act). When we use the criterion that an animal should behave in a way that maximizes its fitness, we don't imagine that an individual animal is performing complex calculations in order to do this. The language of choice and optimisation is used as a convenient short-hand for the action of Natural Selection.

In Example 1.25, the fitnesses for the two behaviours were just the respective reproduction rates r_i. The next example shows that it is not always appropriate to use the number of offspring produced during an animal's life as a measure of fitness.

Example 1.27

Suppose an animal has two possible behaviours:

a_1: Produce 8 offspring, then die. ("Live fast and die young".)

a_2: Produce 5 offspring in the first year, produce 6 more offspring in a second year and then die. ("Live slowly and die old".)

The fitness for behaviour a_1 is simply $\pi(a_1) = 8$. Determining the fitness for behaviour a_2 is a bit more involved. In this case, the sub-population at time t consists of $f(t)$ first-year breeders and $s(t)$ second-year breeders. These numbers change from year to year according to

$$\begin{aligned} f(t+1) &= 5f(t) + 6s(t) \\ s(t+1) &= f(t) \,. \end{aligned}$$

Adding these two equations gives us $\pi(a_2) = 6$. The animal should, therefore, choose a_1. That is, Natural Selection should produce a population of animals which "live fast and die young". This tells us that we should avoid a common misinterpretation of the phrase "survival of the fittest": it is genetic lines, not individuals, which must survive.

In practice, even the expected number of offspring may not be calculated explicitly. Instead some factor that affects fitness is considered, if it can be reasonably assumed that this is the only component of fitness that is affected by the variety of behaviours being considered.

Example 1.28

Suppose an animal chooses actions from a set $\mathbf{A} = \{a_1, a_2\}$ and the animal's probability of survival to the breeding season is S_i if it chooses action a_i. If the animal survives to breed, it has n offspring. The payoff/fitness for adopting a_i is nS_i. Because the factor n is common to the payoffs for all actions, we may consider only the survival probabilities: $\pi(a_i) = S_i$. So $a^* = a_1$ if $S_1 > S_2$ and $a^* = a_2$ if $S_1 < S_2$.

Exercise 1.12

Before migrating to its breeding site, a bird must try to build up its energy reserves x. The bird can choose to forage in any one of three sites, $i = \{1, 2, 3\}$. On each site the bird has a probability λ_i of being eaten by a predator and at the end of the pre-migration foraging period (if it has not been eaten) a bird's reserves will be either *high* or *low* with certain probabilities. The parameters for each site are given in the following table.

Site	1	2	3
λ	0.2	0.1	0.05
$P(x = high)$	0.8	0.6	0.4

The probability of surviving migration is $M_h = 0.9$ if reserves are *high* and $M_l = 0.5$ if reserves are *low*. If a bird survives, it produces a fixed number of offspring during the next breeding season. Which site should the bird choose?[4]

---

[4] Remember, this is just a shorthand way of asking "what is the result of natural selection acting for many years on a population of birds?"

1.5 Optimal Behaviour

Up till now, we have considered the problem of finding an optimal action a^* from a given set $\mathbf{A}$. However, another type of behaviour may be available to an individual: they may randomise. Does this allow an individual to achieve a higher payoff than if they stick to picking an action?

Definition 1.29

We specify a *general behaviour* β by giving the list of probabilities with which each available action is chosen. We denote the probability that action a is chosen by $p(a)$ and
$$\sum_{a \in \mathbf{A}} p(a) = 1 \ .$$
The set of all possible randomising behaviours (for a given problem) will be denoted by $\mathbf{B}$.

The payoff for using a behaviour β is related to the payoffs for the actions in the obvious way. The payoff for using β is given by
$$\pi(\beta) = \sum_{a \in \mathbf{A}} p(a)\pi(a) \ . \tag{1.5}$$
In an uncertain world, we can also define the payoffs
$$\pi(\beta|x) = \sum_{a \in \mathbf{A}} p(a)\pi(a|x)$$
so that
$$\pi(\beta) = \sum_{x \in \mathbf{X}} P(X=x)\pi(\beta|x) \ . \tag{1.6}$$

Exercise 1.13

Show that the payoff for a behaviour β is the same whether we define it via Equation 1.5 or Equation 1.6.

Definition 1.30

An *optimal behaviour* β^* is one for which
$$\pi(\beta^*) \geq \pi(\beta) \qquad \forall \beta \in \mathbf{B} \tag{1.7}$$
or, if we focus on behaviours rather than payoffs,
$$\beta^* \in \underset{\beta \in \mathbf{B}}{\operatorname{argmax}} \pi(\beta) \ . \tag{1.8}$$

Definition 1.31

The *support* of a behaviour β is the set $\mathbf{A}(\beta) \subseteq \mathbf{A}$ of all the actions for which β specifies $p(a) > 0$.

Theorem 1.32

Let β^* be an optimal behaviour with support $\mathbf{A}^*$. Then $\pi(a) = \pi(\beta^*) \ \forall a \in \mathbf{A}^*$.

Proof

If the set $\mathbf{A}^*$ contains only one action, then the theorem is trivially true. Suppose now that the set $\mathbf{A}^*$ contains more than one action. If the theorem is not true, then at least one action gives a higher payoff than $\pi(\beta^*)$. Let a' the action which gives the greatest such payoff. Then

$$\begin{aligned}
\pi(\beta^*) &= \sum_{a \in \mathbf{A}^*} p^*(a)\pi(a) \\
&= \sum_{a \neq a'} p^*(a)\pi(a) + p^*(a')\pi(a') \\
&< \sum_{a \neq a'} p^*(a)\pi(a') + p^*(a')\pi(a') \\
&= \pi(a')
\end{aligned}$$

which contradicts the original assumption that β^* is optimal. $\square$

A consequence of this theorem is that if a randomising behaviour is optimal, then two or more actions are optimal as well. So, randomisation is not necessary to achieve an maximal payoff. However, it may be used as a tie-breaker for choosing between two or more equally acceptable actions.

Exercise 1.14

A firm may make one of three marketing decisions $\{a_1, a_2, a_3\}$. The profit (in millions of pounds) for each decision depends on the state of the economy $X = \{x_1, x_2, x_3\}$ as given in the table below.

	x_1	x_2	x_3
a_1	6	5	3
a_2	3	5	4
a_3	5	9	1

If $P(X = x_1) = \frac{1}{2}$ and $P(X = x_2) = P(X = x_3) = \frac{1}{4}$, find all optimal behaviours.

2
Simple Decision Processes

2.1 Decision Trees

A man hears that his young daughter always takes a nickel when an adult relative offers her a choice between a nickel and a dime. He explains to his daughter, "A dime is twice as valuable as a nickel, so you should always choose the dime". In a rather exasperated tone, his daughter replies "Daddy, but then people will not offer me any money".

This story is an example of a decision process: a sequence of decisions is made, although the process may terminate before all potential decisions have been taken. The story also illustrates two components of what is considered to be *strategic behaviour*. First, immediate rewards are forgone in the expectation of a payback in the future. Second, the behaviour of others is taken into account. It is the former component that is the main subject of this chapter. While the second component may be present in some of the situations we will look at, the behaviour of all individuals other than the one being considered will be taken as fixed. In Part II, we will allow all players to change their behaviour at will.

To represent the problems like the nickel and dime game pictorially, we can draw a *decision tree*. The times at which decisions are made are shown as small, filled circles. Leading away from these *decision nodes* is a branch for every action that could be taken at that node. When every decision has been made, one reaches the end of one path through the tree. At that point, the payoff for following that path is written. The convention we will follow is for time to increase as one goes down the page, so the tree is drawn "upside-down".

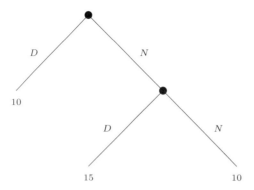

Figure 2.1 Choosing a nickel (N) or a dime (D) on (at most) two occasions. The payoff in cents is given at the end of each branch of the tree.

Example 2.1

Suppose that the adult will offer the "nickel or dime" choice at most twice: if the girl takes the dime on the first occasion, then the choice will be offered only once. The nickel and dime problem can then be represented by the tree shown in Figure 2.1. If she chooses a dime (action D) at the first opportunity, then she receives ten cents and no further offer is made. On the other hand, if she chooses the nickel (action N), she gets five cents and a second choice. It is clear what the girl should do. If she chooses the nickel the first time and then the dime, she gets a payoff of fifteen cents; if she follows any other course of action, she gets only ten cents. Therefore, she should choose the nickel first and then the dime.

2.2 Strategic Behaviour

The word "strategy" is derived from the Greek word *strategos* ($\sigma\tau\rho\alpha\tau\epsilon\gamma o\varsigma$) meaning "military commander" and, colloquially, a strategy is a plan of action.

Definition 2.2

A *strategy* is a rule for choosing an action at every point that a decision might have to be made. A *pure strategy* is one in which there is no randomisation. The set of all possible pure strategies will be denoted **S**.

2.2 Strategic Behaviour

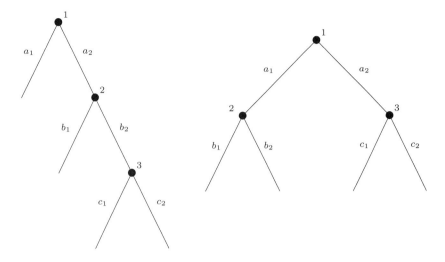

Figure 2.2 Two decision trees that could have the pure strategy set given in Example 2.3.

Suppose that there are n decision nodes and that at each decision node i there is an action set $\mathbf{A}_i$ describing the choices that can be made at that point. Some or all of the sets $\mathbf{A}_i$ may be identical. Then the set of pure strategies $\mathbf{S}$ is given by the cross-product of all the action sets: $\mathbf{S} = \mathbf{A}_1 \times \mathbf{A}_2 \times \cdots \times \mathbf{A}_n$.

Example 2.3

Suppose there are three decision nodes at which the action sets are $\mathbf{A} = \{a_1, a_2\}$, $\mathbf{B} = \{b_1, b_2\}$ and $\mathbf{C} = \{c_1, c_2\}$. Then the set of pure strategies is given by the set of eight triples

$$\mathbf{S} = \{a_1 b_1 c_1, a_1 b_1 c_2, a_1 b_2 c_1, a_1 b_2 c_2, a_2 b_1 c_1, a_2 b_1 c_2, a_2 b_2 c_1, a_2 b_2 c_2\}.$$

This strategy set could apply to either of the decision trees illustrated in Figure 2.2.

Definition 2.4

The observed behaviour of an individual following a given strategy is called the *outcome* of the strategy.

The definition of a strategy leads to some redundancy in terms of outcomes. A pure strategy picks a path through the decision tree from the initial point to one of the terminal points. However, a pure strategy is not just a path through the decision tree: a pure strategy specifies the action that would be taken at every decision node including those that will not be reached if the strategy is followed. In other words, the observed behaviour of an individual only provides us with part of the strategy itself.

Example 2.5

Consider the "nickel or dime" game shown in Figure 2.1. The pure strategy set is $\mathbf{S} = \{NN, ND, DN, DD\}$, where each pair of actions represents the choices made in the natural (time-increasing) order. Two of these strategies, DN and DD, yield the same outcome because choosing the dime at the first decision node means that no further decisions have to be made.

Because strategies that give the same outcome lead to the same payoff, it is sometimes useful to introduce the concept of a "reduced" set of pure strategies, which removes this redundancy from the discussion.

Definition 2.6

A *reduced strategy set* is the set formed when all pure strategies that lead to indistinguishable outcomes are combined.

Example 2.7

For the "nickel or dime" game shown in Figure 2.1. The reduced strategy set is $\mathbf{S}_R = \{NN, ND, DX\}$, where the combination DX means "dime at the first decision node and anything at the other decision node".

Exercise 2.1

(a) Consider a variant of the "nickel or dime" game from Example 2.1 where the child is offered nickels or dimes on three occasions at most. Draw the tree for this decision problem, determine the pure-strategy set and find the optimal strategy?

(b) Suppose the child is offered the nickel or dime choice on n occasions. What is the optimal strategy?

(c) Suppose the adult offers the child a choice between a nickel or a dime. If the child takes the dime, then the game stops. If the child takes the nickel, then the choice is offered again with probability p. If $p < 1$, then the game will eventually terminate, perhaps because the adult gets bored. What is the child's optimal strategy?

2.3 Randomising Strategies

When there is only a single decision to be made, the sets of actions and pure strategies are identical. There is also only one way of specifying randomising behaviour.

Example 2.8

Suppose the action (or pure strategy) set is $\{a_1, a_2\}$. A general behaviour specifies using a_1 with probability p and a_2 with probability $1-p$. In Section 1.5, we denoted this by $\beta = (p, 1-p)$.

When there is (potentially) more than one decision to be made, the action sets and pure strategy sets are no longer identical and there are now two, conceptually different ways of representing a randomizing behaviour. To distinguish between them we shall call one a "mixed strategy" and the other a "behavioural strategy".

Definition 2.9

A *mixed strategy* σ specifies the probability $p(s)$ with which each of the pure strategies $s \in \mathbf{S}$ is used.

Suppose the set of strategies is $\mathbf{S} = \{s_a, s_b, s_c, \ldots\}$, then a mixed strategy can be represented as a vector of probabilities:

$$\sigma = (p(s_a), p(s_b), p(s_c), \ldots) \ .$$

A pure strategy can then be represented as a vector where all the entries are zero except one. For example,

$$s_b = (0, 1, 0, \ldots) \ .$$

Mixed strategies can, therefore, be represented as linear combinations of pure strategies:

$$\sigma = \sum_{s \in \mathbf{S}} p(s) s \ .$$

Remark 2.10

Often these linear combinations are written symbolically. For example, in the "nickel or dime" game, the mixed strategy in which NN is used with probability $\frac{1}{4}$ and DN is used with probability $\frac{3}{4}$ might be written as

$$\sigma = \frac{1}{4} NN + \frac{3}{4} DN \ .$$

Definition 2.11

The *support* of a mixed strategy σ is the set $\mathbf{S}(\sigma) \subseteq \mathbf{S}$ of all the pure strategies for which σ specifies $p(s) > 0$.

For a mixed strategy, the randomisation takes place once *before* the decision tree is traversed: once a strategy has been chosen, the path through the tree is fixed.

Definition 2.12

Let the decision nodes be labelled by an indicator set $\mathbf{I} = \{1, 2, 3, \ldots n\}$. At node i, the action set is $\mathbf{A}_i = \{a_1^i, a_2^i, \ldots, a_{k_i}^i\}$ (where we have allowed the number of available actions k_i to be different at each decision node i). An individual's behaviour at node i is determined by a probability vector $\mathbf{p}_i$ where $\mathbf{p}_i = (p(a_1^i), p(a_2^i), \ldots, p(a_{k_i}^i))$ and $p(a_j^i)$ is the probability with which he selects action $a_j^i \in \mathbf{A}_i$ (if, in fact, they reach decision node i). A *behavioural strategy* β is the collection of probability vectors

$$\beta = \{\mathbf{p}_1, \mathbf{p}_2, \ldots, \mathbf{p}_n\} \ . \tag{2.1}$$

In contrast to a mixed strategy, a behavioural strategy causes randomisation to take place several times *as* the decision tree is traversed.

As we shall see, these two representations of randomising behaviour are interchangeable in the sense that every mixed strategy has an equivalent behavioural representation and every behavioural strategy has an equivalent mixed representation. In each case, a strategy defined in terms of one type of representation may have more than one equivalent strategy defined in terms of the other representation. Before we give a proper definition, let's look at the idea of equivalence by means of an example that illustrates two strategies that are *not* equivalent.

Example 2.13

Consider the "nickel or dime" game shown in figure 2.1. One mixed strategy for this decision process is $\sigma = \frac{1}{2}NN + \frac{1}{2}DD$. It would be tempting to believe that a behavioural equivalent to this strategy is $\beta = \left(\left(\frac{1}{2}, \frac{1}{2}\right), \left(\frac{1}{2}, \frac{1}{2}\right)\right)$, but this would be incorrect. To see why, note that there are three paths through the decision tree. Let's call them "dime only", "all nickels" and "nickel then dime". The mixed strategy σ picks out the paths "dime only" and "all nickels" each with a probability $\frac{1}{2}$ and picks "nickel then dime" with probability zero. However, the behavioural strategy β specifies choosing the action D at the later decision node with probability $\frac{1}{2}$. Therefore, the path "nickel then dime" would be picked with probability $\frac{1}{4}$ and not zero. The strategies σ and β are, therefore, not equivalent.

Definition 2.14

A behavioural strategy and a mixed strategy are equivalent if they assign the same probabilities to each of the possible pure strategies that are available.

It follows immediately that equivalent mixed and behavioural strategies have the equal payoffs.

Example 2.15

A behavioural strategy which is equivalent to the mixed strategy σ in example 2.13 is $\beta = \left(\left(\frac{1}{2}, \frac{1}{2}\right), (1, 0)\right)$. Furthermore, any of the mixed strategies

$$\sigma_x = \frac{1}{2}NN + \left(\frac{1}{2} - x\right)DD + xDN \quad \text{with} \quad x \in \left[0, \frac{1}{2}\right]$$

is equivalent to the behavioural strategy $\beta = \left(\left(\frac{1}{2}, \frac{1}{2}\right), (1, 0)\right)$.

Exercise 2.2

Show that the following behavioural and mixed strategies for the "nickel or dime" game of Example 2.1 all have the same payoff.

$$\beta = \left(\left(\frac{1}{2}, \frac{1}{2}\right), (1, 0)\right)$$

$$\sigma = \frac{1}{2}ND + \left(\frac{1}{2} - x\right)DD + xDN \quad \text{with} \quad x \in \left[0, \frac{1}{2}\right]$$

Theorem 2.16

(a) Every behavioural strategy has a mixed representation and (b) every mixed strategy has a behavioural representation.

Proof

(a) An individual using a pure strategy $s \in \mathbf{S}$ will pass through a set of decision nodes $\mathbf{I}(s) \subseteq \mathbf{I}$, choosing some action $a^i(s) \in \mathbf{A}_i$ for each $i \in \mathbf{I}(s)$. At each decision node $i \in \mathbf{I}$, a given behavioural strategy β would prescribe choosing that action with probability $p(a^i(s))$. So the probability that an individual using β would traverse the decision tree via the decision nodes $\mathbf{I}(s)$ is

$$p(s) = \prod_{i \in \mathbf{I}(s)} p(a^i(s)) .$$

A mixed strategy representation of β is then

$$\sigma_\beta = \sum_{s \in \mathbf{S}} p(s) s$$

because

$$\sum_{s \in \mathbf{S}} \prod_{i \in \mathbf{I}(s)} p(a^i(s)) = 1 .$$

(b) Let $\sigma = \sum_{s \in \mathbf{S}} p(s) s$ be some mixed strategy. For each pure strategy s, let $\mathbf{I}(s) \subseteq \mathbf{I}$ be the set of decision nodes an individual encounters when he follows strategy s. For each decision node $i \in \mathbf{I}$, let $\mathbf{S}(i) \subseteq \mathbf{S}$ be the set of pure strategies that reach decision node i. Then the probability that an individual following the mixed strategy σ will reach decision node i is

$$p_\sigma(i) = \sum_{s \in \mathbf{S}(i)} p(s) .$$

Let $\mathbf{S}(a^i, i) \subseteq \mathbf{S}(i)$ be the set of pure strategies that reach decision node i and choose action $a^i \in \mathbf{A}_i$ at that point. Then the probability that an individual following the mixed strategy σ will reach decision node i and choose a^i is

$$p_\sigma(a^i, i) = \sum_{s \in \mathbf{S}(a^i, i)} p(s) .$$

Provided $p_\sigma(i) \neq 0$ we can define the probability of choosing a^i at i as

$$p(a^i) = \frac{p_\sigma(a^i, i)}{p_\sigma(i)} .$$

If $p_\sigma(i) = 0$ for some decision node i then *any* set of probabilities $p(a^i)$ with $\sum_{a^i \in \mathbf{A}_i} p(a^i) = 1$ will suffice. Clearly $\sum_{a^i \in \mathbf{A}_i} p(a^i) = 1 \; \forall i \in \mathbf{I}$, so the collection

2.4 Optimal Strategies

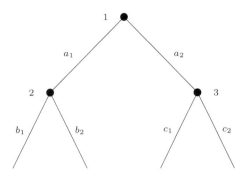

Figure 2.3 Decision tree for Exercise 2.3.

of these probabilities for all actions at all decision nodes forms a behavioural representation of the mixed strategy σ. □

It follows from this theorem that we are free to choose the representation for strategies that best suits the problem in hand.

Exercise 2.3

Consider the decision tree shown in Figure 2.3. Find the all behavioural strategy equivalents for the mixed strategies (a) $\sigma = \frac{1}{2}a_1b_1c_1 + \frac{1}{2}a_2b_2c_2$ and (b) $\sigma = \frac{1}{3}a_1b_1c_1 + \frac{1}{3}a_1b_2c_1 + \frac{1}{3}a_1b_1c_2$.

2.4 Optimal Strategies

In Chapter 1, we saw that randomising behaviour was not required for single decisions, in the sense that an optimal action could always be found. A similar result holds for decision processes.

Lemma 2.17

Let σ^* be an optimal mixed strategy with support $\mathbf{S}^*$. Then $\pi(s) = \pi(\sigma^*)$ $\forall s \in \mathbf{S}^*$.

Proof

If the set $\mathbf{S}^*$ contains only one strategy, then the theorem is trivially true. Suppose now that the set $\mathbf{S}^*$ contains more than one strategy. If the theorem is not true, then at least one strategy gives a higher payoff than $\pi(\sigma^*)$. Let s' be the strategy that gives the greatest such payoff. Then

$$\begin{aligned}
\pi(\sigma^*) &= \sum_{s \in \mathbf{S}^*} p^*(s)\pi(s) \\
&= \sum_{s \neq s'} p^*(s)\pi(s) + p^*(s')\pi(s') \\
&< \sum_{s \neq s'} p^*(s)\pi(s') + p^*(s')\pi(s') \\
&= \pi(s')
\end{aligned}$$

which contradicts the original assumption that σ^* is optimal. $\square$

Theorem 2.18

For any decision process, an optimal pure strategy can always be found.

Proof

From Theorem 2.16, we know that every behavioural strategy has at least one equivalent mixed strategy. It follows that no behavioural strategy can have a payoff greater than that which could be achieved by using a mixed strategy. It, therefore, follows from the preceding lemma that, if an optimal mixed strategy exists, then an optimal pure strategy also exists. $\square$

So far, we have adopted a simple procedure for finding a optimal strategy: list the possible pure strategies, calculate the payoff for each of these, and pick one that gives the optimal payoff. However, the burden of the procedure increases exponentially as the decision tree becomes larger. A tree with n decision nodes each with 2 possible actions leads to 2^n pure strategies. Fortunately, we can reduce this burden by employing the *Principle of Optimality*. This principle states that from any point on an optimal path, the remaining path is optimal for the decision problem that starts at that point. In other words, to find the optimal decision *now*, we should assume that we will behave optimally in the *future*.

2.4 Optimal Strategies

Definition 2.19

A *partial history* h is the sequence of decisions that have been made by an individual up to some specified time. At the start of a decision process (when no decisions have been made), we have the *null history*, $h = \emptyset$. A *full history* for a strategy s is the complete sequence of all decisions that would be made by an individual following s and will be denoted $H(s)$.

Remark 2.20

If an individual has perfect recall (i.e., remembers all their past decisions), then each decision node has a unique history and each history specifies a unique (current) decision node.

Define the subset of pure strategies $S(h) \in S$ that contains all the strategies with history h but that differ in the actions taken in the future. Then the optimal payoff an individual can achieve given that they have history h is

$$\pi^*(s|h) = \max_{s \in S(h)} \pi(s).$$

Assume that the individual now has a choice from a set of actions $A(h)$. After that decision has been made, the history will be the sequence h with the chosen action a appended. We will write this as h, a.

Theorem 2.21 (The Optimality Principle)

For an individual with perfect recall:

1. $\pi^*(s|H(s)) = \pi(s)$
2. $\pi^*(s|h) = \max_{a \in A(h)} \pi^*(s|h, a)$
3. $\pi^* = \max_{s \in S(\emptyset)} \pi^*(s|\emptyset)$

Proof

1. By the definition of $H(s)$, the individual has no more decisions to make and the best payoff they can get is the payoff they have already achieved by following the strategy s.

2. A pure strategy is a sequence of actions $\{a_0, a_1, \ldots, a_h, a_{h+1}, a_{h+2}, \ldots, a_H\}$ so
$$\pi(s) = \pi(a_0, a_1, \ldots, a_h, a_{h+1}, a_{h+2}, \ldots, a_H).$$

Let the partial history h be a given sequence $\{a_0, a_1, \ldots, a_h\}$ then

$$\pi^*(s|h) = \max_{a_{h+1}} \max_{a_{h+2}} \ldots \max_{a_H} \pi(a_0, a_1, \ldots, a_h, a_{h+1}, a_{h+2}, \ldots, a_H)$$
$$= \max_{a_{h+1}} \pi^*(s|h, a_{h+1}).$$

3. The history $h = \emptyset$ denotes optimisation problem starting from the very beginning. So $S(\emptyset) = S$ and

$$\max_{s \in S(\emptyset)} \pi^*(s|\emptyset) = \max_{s \in S} \pi(s)$$
$$= \pi^*.$$

$\square$

The Optimality Principle leads directly to a convenient method for solving dynamic decision problems. If we wish to find the optimal decision now by assuming that we will behave optimally in the future, it makes sense to sort out the future decisions first. In other words, we should work *backwards* through the decision tree – a procedure known as *backward induction*.

Example 2.22

Consider the decision tree shown in Figure 2.4. To work backwards through this tree, we start at either decision node 2 or decision node 3. It does not matter which: all that matters is that no decision node is considered before all the decision nodes that follow on from it have been dealt with. At decision node 3 we would choose C (rather than D), which gives us a payoff of 8. At decision node 2 we would choose D to get a payoff of 7. Now consider decision node 1. Assuming that we will choose optimally in the future *whatever we do now*, choosing A leads to a final payoff of 7 whereas choosing B leads to a payoff of 8. The optimal strategy is therefore BDC (in the order of the labelling of the decision nodes).

The previous example shows, at least, that backward induction produces the same result as a complete strategic analysis. However, the advantages of the approach seem fairly minimal. The real power of backward induction reveals itself when we consider problems for which drawing a complete decision tree is impractical if not actually impossible. Such problems are considered in the next chapter.

Exercise 2.4

Consider a female bird choosing a mate from three displaying males. The attributes of the males are summarised by the following table.

2.4 Optimal Strategies

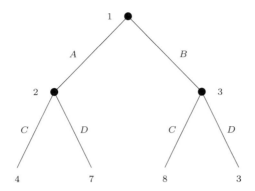

Figure 2.4 Decision tree for Example 2.22.

Male	Genetic quality	Cares for chicks?
1	High	No
2	Medium	Yes
3	Low	Yes

Suppose that the value of offspring depends on the genetic quality of the father. The value of offspring is v_H, v_M, and v_L for the males of high, medium, and low quality, respectively, with $v_H > v_M > v_L$. Once she has mated, the female can choose to care for the chicks or desert them. Chicks that are cared for by both parents will certainly survive; those cared for by only one parent (of either sex) have a 50% chance of survival; and those deserted by both parents will certainly die. Draw the decision tree and find the female's optimal strategy.

3
Markov Decision Processes

3.1 State-dependent Decision Processes

In this chapter, we add an extra layer of complexity to our models of decision making by introducing the idea of a state-dependent decision process. The processes we will consider can either be deterministic or stochastic. To begin with, we will assume that the process must terminate by an a priori fixed time T (a "finite horizon" model). In principle, decisions can be made at times $t = 0, 1, 2, \ldots, T-1$, although the actual number of decisions made may be fewer than T if the process terminates early as a consequence of the actions taken. Models that have no a priori restriction on the number of decisions to be taken ("infinite horizon" models) will be considered in the next chapter.

We consider individuals to have some state variable x taken from a set **X** that may be either discrete or continuous. This state could represent an individual's wealth, number of offspring, need for food, etc. We allow individuals to condition their behaviour on the state they find themselves in at any given time, and we will denote the set of actions available in state x at time t by $\mathbf{A}(x,t)$. The action taken causes a transition to a new state: that is, at time t, the action a_t induces a transition $x_t \to x_{t+1}$. Thus we are considering a deterministic process that consists of the sequence of pairs (x_t, a_t) for $t = 0, 1, 2, \ldots$.

Definition 3.1

An alternating sequence of states and actions $x_0, a_0, x_1, a_1, x_2, \ldots$ is called the *history* of a process. We will denote by $h_t = (x_0, a_0, x_1, a_1, \ldots, x_t)$ the history of the process up to time t.

How are decisions to be made? In its simplest form, a pure strategy s could just be taken as a sequence of actions $a_0, a_1, a_2, \ldots$. More generally, we can allow strategies to specify a choice of action $a(x, t)$ for each state and each decision time. Starting from x_0, this strategy generates a history $x_0, a_0, x_1, a_1, x_2, \ldots$ where we have written $a_t = a(x_t, t)$. We will assume that this history generates a sequence of rewards $r_t(x_t, a_t)$ for $t = 1, 2, \ldots, T - 1$. That is, an individual in state x_t at time t who uses action a_t will receive an immediate reward of $r_t(x_t, a_t)$. For finite processes, we also include an optional terminal reward $r_T(x_T)$ that is received at the end of the process. The total reward obtained starting in state x if an individual follows a strategy s is given by

$$\pi(x|s) = \sum_{t=0}^{T-1} r_t(x_t, a_t) + r_T(x_T).$$

In some problems, the process may start in a known state x_0. In which case, we only have to consider one payoff, namely $\pi(x_0|s)$.

We have so far assumed that the state transition caused by choosing action a is deterministic. We will now consider stochastic decision processes in which the state at time t is a random variables, which we denote by X_t. The probabilities for the state transition $x_t \to x_{t+1}$ can, in general, depend on the whole history of the process as well as the action chosen a_t

$$P(X_{t+1} = x_{t+1}) = p(x_{t+1}|h_t, a_t)$$

with

$$\sum_{x_{t+1} \in \mathbf{X}} p(x_{t+1}|h_t, a_t) = 1.$$

for all times, histories and actions. We should also consider randomising strategies σ such that the action chosen is also a random variable A_t. The total reward obtained starting in state x if an individual follows a strategy σ is given by

$$\begin{aligned}\pi(x|\sigma) &= \mathbb{E}\left[\sum_{t=0}^{T-1} r_t(X_t, A_t) + r_T(X_T)\right] \\ &= \sum_{t=0}^{T-1} \mathbb{E}\left[r_t(X_t, A_t)\right] + \mathbb{E}\left[r_T(X_T)\right].\end{aligned}$$

We will use the notation $\pi^*(x)$ for the payoff obtained by following the optimal strategy starting from state x. That is,

$$\pi^*(x) \equiv \pi(x|\sigma^*) \ .$$

An individual's aim is to maximise this *expected* total reward.

3.2 Markov Decision Processes

There is a special class of decision processes in which the state transition probabilities depend only upon the current state and not on how that state was reached.

Definition 3.2

A decision process is said to have the *Markov Property* if $p(x_{t+1}|h_t, a_t) = p(x_{t+1}|x_t, a_t)$.

Definition 3.3

A decision process with the Markov property is called a *Markov Decision Process* (MDP) (named in honour of the Russian mathematician Andrei Andreevich Markov).

From now on, we will consider only MDPs and not more general decision processes. Clearly, it is not an easy task to compute the payoff for a general strategy and hence find an optimal one. But for finite horizon MDPs the method of backward induction (the *Principle of Optimality* – see Section 2.4) comes to our rescue.

We begin our discussion of Markov Decision Processes by considering a simple example. The example is deterministic and can be solved using the Lagrangian method for constrained optimisation (see Appendix A). We will show that the same optimal strategy can also be found by backward induction (dynamic programming).

Example 3.4

Consider an investor making a sequence of decisions about how much of their current capital to consume (i.e., spend on goods) and how much to invest. That is, at time t the investor's capital x_t is reduced by an amount c_t and the remainder is invested at an interest rate $r-1$ to produce an amount of capital $r(x_t - c_t)$ at the next decision time. For simplicity, let us restrict ourselves to the two-period problem of an investor who starts with known capital x_0 and makes decisions about consumption at $t = 0$ and $t = 1$. We assume that the investor only gains immediate benefit from consumption (the only reason for investment is to obtain the benefit of consumption in the future). We shall also assume that the investor's utility for consumption is logarithmic, i.e. $\pi(c_0, c_1) = \ln(c_0) + \ln(c_1)$.

First we solve this problem using the Lagrangian method. The state equation is $x_1 = r(x_0 - c_0)$ and we must have $c_1 \leq x_1$ so the constraint equation is $c_1 \leq r(x_0 - c_0)$. Therefore, the Lagrangian is $L(c_0, c_1) = \ln(c_0) + \ln(c_1) - \lambda(c_1 + rc_0 - rx_0)$ and we must simultaneously satisfy the following three equations.

$$\frac{1}{c_0^*} - \lambda r = 0$$

$$\frac{1}{c_1^*} - \lambda = 0$$

$$c_1^* + rc_0^* - rx_0 = 0$$

Solving the first pair of equations provides the following relation between the consumptions during the two periods: $c_1^* = rc_0^*$. Substituting this back into the constraint equation gives the optimal strategy as

$$c_0^* = \frac{1}{2}x_0 \quad \text{and} \quad c_1^* = \frac{1}{2}rx_0$$

To solve the same problem again by the method of backward induction we proceed as follows. At $t = 1$ the payoff is $\ln(c_1)$ subject to the constraint $c_1 \leq x_1$ where x_1 is the amount of capital that the investor has at this time. The optimal decision is, therefore, $c_1^*(x_1) = x_1$. Note that we don't know the value of x_1 because it depends on the behaviour at $t = 0$ through the state equation $x_1 = r(x_0 - c_0)$. So what we have is an optimal decision for *any* value of x_1. At $t = 0$, the investor's problem is to maximise the total payoff assuming optimal behaviour at $t = 1$, i.e., find

$$c_0^* \in \underset{c_0 \in [0, x_0]}{\operatorname{argmax}} \left(\ln(c_0) + \ln(x_1(c_0)) \right). \tag{3.1}$$

Differentiating the payoff and setting the result to zero gives $x_1 - rc_0^* = 0$. Substituting for x_1 using the state equation gives $c_0^* = \frac{1}{2}x_0$. (We could write this as $c_0^*(x_0)$ but because x_0 is assumed to be known, we can drop this dependence.) Thus the optimal strategy can be written as $s^*(x, t) = (c_0^*, c_1^*(x_1))$ with

$$c_0^* = \frac{1}{2}x_0 \quad \text{and} \quad c_1^*(x_1) = x_1.$$

Although this solution looks slightly different from the one found by the Lagrangian method it is, in fact, identical because

$$\begin{aligned} c_1^*(x_1) &= r(x_0 - c_0^*) \\ &= \frac{1}{2}rx_0. \end{aligned}$$

Exercise 3.1

Consider a three-period consumption and investment model with logarithmic utility for consumption. Apart from the change to three periods, make the same assumptions that were used in Example 3.4.

(a) Find the optimal consumption strategy using the Lagrangian method. [HINT. We can rewrite the constraint for the two-period problem in the form of total consumption discounted to initial (period 0) value

$$c_0 + \frac{c_1}{r} \leq x_0.$$

You may find it useful to write the constraint for the current problem in this way.]

(b) Solve the model by backward induction and show that the solution is identical to the one obtained using the Lagrangian method.

Backward induction in a state-dependent problem is often called "dynamic programming". Equation 3.1 is an example of a *dynamic programming equation*, and it is just the second equation from Theorem 2.21 written in a way that is suitable for state-dependent decision processes. As a prelude to the introduction of stochastic dynamic programming we will now develop a general form for the dynamic programming equation in the deterministic case.

The basis of dynamic programming is answering the following question. *What is the best action now, assuming optimal behaviour at all potential future decision points?* The word "potential" is included to indicate that we have to know what would be done in all future states, including those that may not be reached once the optimal decision has been found and taken. Without that information, we could not decide what is optimal. So, let us define $\pi_t^*(x)$ to be the future payoff for starting in state x at time t providing we behave optimally for times $t, t+1, t+2, \ldots, T-1$. That is,

$$\pi_t^*(x) = \sum_{\tau=t}^{T-1} r_t(x_\tau^*, a_\tau^*) + r_T(x_T^*) \qquad (3.2)$$

where $x_t = x$ and the sequence of states $(x_\tau^*)_{\tau=t+1,\ldots,T}$ is generated by following the sequence of optimal actions $(a_\tau^*)_{\tau=t,\ldots,T-1}$. We can then write the general form of the deterministic dynamic programming equation as

$$\pi_t^*(x) = \max_{a \in \mathbf{A}(x)} \left[r_t(x,a) + \pi_{t+1}^* \left(x_{t+1}(a) \right) \right] \quad (3.3)$$

where $x_{t+1}(a)$ is the state reached from x by using an action a taken from the set of actions available in state x, $\mathbf{A}(x)$. The backward induction process is started by setting

$$\pi_T^*(x) = r_T(x_T) \quad \forall x \in \mathbf{X}$$

and the payoff achieved by the optimal strategy s^* for the starting state(s) of interest x is given by $\pi(x|s^*) = \pi_0^*(x)$.

Exercise 3.2

Relate the various elements of Example 3.4 to the elements in the general description of dynamic programming.

3.3 Stochastic Markov Decision Processes

In a deterministic MDP, the choice of action in a particular state uniquely determines the state of the process at the next decision point. (In some applications, therefore, the actions are conveniently described in terms of choosing the next state.) For the rest of this chapter, we will consider decision processes in which the transitions between states may be uncertain. We will assume that these transition probabilities are time-independent: given that an individual is in state x at time t and chooses action a, the probability that they find themselves in state x' at time $t+1$ is $p(x'|x,a) \leq 1$, $\forall x' \in \mathbf{X}$. Although we have made this "stationarity" assumption, the optimal strategy may nevertheless be time-dependent. We have seen this already in Example 3.4 – a deterministic MDP is, after all, just a stochastic MDP where all the transition probabilities happen to be either 0 or 1. That example also clearly illustrates the fact that the time dependence can arise from the finiteness of the problem, because we have a zero terminal reward in every state.

Under certain conditions, a stochastic MDP has a simple diagrammatic representation. These conditions are as follows: the number of states is small; the number of actions available in each state is small and independent of time (at least for $t < T$); and the rewards obtained for each state-action pair are independent of time ($r_t(x,a) = r(x,a)$ for $t < T$). This diagrammatic representation is best introduced by means of an example.

3.3 Stochastic Markov Decision Processes

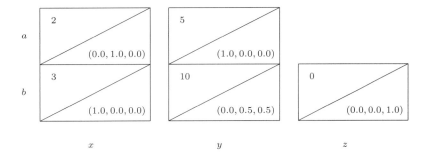

Figure 3.1 Diagrammatic representation of the stochastic MDP described in Example 3.5.

Example 3.5

The set of states is $\mathbf{X} = \{x, y, z\}$. In states x and y, we can choose an action from the sets $\mathbf{A}(x) = \mathbf{A}(y) = \{a, b\}$, and in state z we have the single-choice set $\mathbf{A}(z) = \{b\}$.[1] If we choose action a in state x, then we receive a reward $r(x, a) = 2$ and move to state y with probability 1. If we choose action b in state x, then we receive a reward $r(x, b) = 3$ and remain in state x with probability 1. In state y, if we choose a, then we receive $r(y, a) = 5$ and move to state x with probability 1, whereas choosing b gives us $r(y, b) = 10$ and we transfer to state z with 50% probability and remain in state y with 50% probability. If we find ourselves in state z, then we can only choose b, which gives us $r(z, b) = 0$ and we remain in state z with probability 1.

This lengthy description can be presented much more concisely by means of the diagram shown in Figure 3.1. In the case of a finite horizon problem this diagrammatic description must be supplemented by specifying the horizon T and the final rewards $r_T(x)$, $\forall x \in \mathbf{X}$.

Remark 3.6

Note that in Figure 3.1 the state z is *absorbing*: if the process ever arrives in state z it stays there. Furthermore, the payoff received in state z is zero. The existence of a zero-payoff, absorbing state is quite common in MDP models.

---

[1] A choice without an alternative is often known as *Hobson's choice*, though the phrase is also applied to "take it or leave it" choices, which would be a two-element set. The term is said to originate from Thomas Hobson (ca. 1544–1631) who owned a livery stable at Cambridge, England. He allegedly required every customer to take either the horse nearest the stable door or none at all.

For example, in a biological context, a transition to state z could represent the death of the organism.

To solve problems such as that given in Example 3.5, we will need a stochastic version of the dynamic programming equation (Equation 3.3). Fortunately, this is easy to write down. Define $\pi_t(x|a)$ to be the future payoff for choosing action a in state x at time t and behaving optimally for times $t+1, t+2, \ldots, T-1$. That is,
$$\pi_t(x|a) = r(x,a) + \sum_{x' \in \mathbf{X}} p(x'|x,a)\, \pi^*_{t+1}(x') \ .$$
Now define $\pi^*_t(x)$ to be the future payoff for starting in state x at time t providing we behave optimally for times $t, t+1, t+2, \ldots, T-1$. (As before, $\pi^*_t(x)$ is given by Equation 3.2.) The *stochastic dynamic programming equation* is then
$$\pi^*_t(x) = \max_{a \in A} \pi_t(a|x) \tag{3.4}$$
$$= \max_{a \in A} \left[r_t(x,a) + \sum_{x' \in X} p(x'|x,a)\, \pi^*_{t+1}(x') \right]. \tag{3.5}$$

If two actions, say a and b, both lead to the maximum future payoff $\pi^*(x)$, then either can be chosen. (In fact, the combination "a with probability p and b with probability $1-p$" also gives the same maximum future payoff, but in this case the randomisation is best regarded as a way of breaking the tie rather than as a necessity. See Theorems 1.32 and 2.18.)

Except in very rare cases, stochastic MDPs are not solved for arbitrary parameter values. Take the problem shown in Figure 3.1, for example. If all the rewards, transition probabilities, and the horizon T were left unspecified, there would be 15 parameters to deal with. Even if an analytic solution could be found (which would be difficult), understanding the way a solution changes as all these parameters are varied is not really feasible. Instead, the usual approach is to fix all of the parameter values and find a solution numerically. Often a computer program is employed; but if the number of states, the number of actions and the time horizon are all small it is possible to do this "by hand".

Example 3.7

Consider the MDP shown in Figure 3.1. We will additionally assume that decisions are to be made at times $t = 0, 1$ and 2 (i.e., $T = 3$) and that $r_3(x) = r_3(y) = r_3(z) = 0$.

The first thing to note is that in state z, there is no choice to be made: $a^*(z,t) = b$ and $\pi^*_t(z) = 0$, $\forall t$. The absence of a terminal reward $r_T(x) = 0$ also gives us $\pi^*_3(x) = \pi^*_3(y) = 0$.

3.3 Stochastic Markov Decision Processes

At time $t = 2$ in state x, we have

$$\pi_2(x|a) = 2 + \pi_3^*(y) = 2$$
$$\pi_2(x|b) = 3 + \pi_3^*(x) = 3.$$

So $a^*(x, 2) = b$ and $\pi_2^*(x) = 3$. In state y, we have

$$\pi_2(y|a) = 5 + \pi_3^*(x) = 5$$
$$\pi_2(y|b) = 10 + \frac{1}{2}\pi_3^*(y) + \frac{1}{2}\pi_3^*(z) = 10.$$

So $a^*(y, 2) = b$ and $\pi_2^*(y) = 10$.

At time $t = 1$ in state x, we have

$$\pi_1(x|a) = 2 + \pi_2^*(y) = 12$$
$$\pi_1(x|b) = 3 + \pi_2^*(x) = 6.$$

So $a^*(x, 1) = a$ and $\pi_1^*(x) = 12$. In state y, we have

$$\pi_1(y|a) = 5 + \pi_2^*(x) = 8$$
$$\pi_1(y|b) = 10 + \frac{1}{2}\pi_2^*(y) + \frac{1}{2}\pi_2^*(z) = 15.$$

So $a^*(y, 2) = b$ and $\pi_1^*(y) = 15$.

At time $t = 0$ in state x, we have

$$\pi_0(x|a) = 2 + \pi_1^*(y) = 17$$
$$\pi_0(x|b) = 3 + \pi_1^*(x) = 15.$$

So $a^*(x, 0) = a$ and $\pi_0^*(x) = 17$. In state y, we have

$$\pi_0(y|a) = 5 + \pi_1^*(x) = 17$$
$$\pi_0(y|b) = 10 + \frac{1}{2}\pi_1^*(y) + \frac{1}{2}\pi_1^*(z) = 17.5.$$

So $a^*(y, 2) = b$ and $\pi_0^*(y) = 17.5$.

The solution of the problem is, therefore, the optimal strategy[2]

$$s^* = \begin{matrix} \\ x \\ y \\ z \end{matrix} \begin{pmatrix} t=0 & t=1 & t=2 \\ a & a & b \\ b & b & b \\ b & b & b \end{pmatrix} \qquad (3.6)$$

and a payoff of 17 if the process starts in state x or a payoff or 17.5 if the process starts in state y.

---
[2] At least, it is the optimal *backward induction* strategy. We will see later that it is, in fact, optimal in the sense that it is the best of all possible strategies.

Exercise 3.3

Use the strategy in Equation 3.6 to follow the decision process forward in time (i.e., starting at $t = 0$ then moving to $t = 1$, etc.). Check that the strategy, indeed, produces the expected payoffs $\pi_0^*(x)$ and $\pi_0^*(y)$ that were found by backward induction.

3.4 Optimal Strategies for Finite Processes

We now turn to the question of whether the pure strategy found by dynamic programming is truly optimal. It turns out that the strategy found by dynamic programming is still optimal when the class of possible strategies is enlarged to include more general types of strategy – while other strategies may do as well as the dynamic programming strategy, none can do better. Let us first consider the possibility of randomising strategies that depend only upon the current state.

Definition 3.8

Denote the set of actions available in state x at time t by $\mathbf{A}(x,t)$. A general *Markov strategy* β specifies using action $a \in \mathbf{A}(x,t)$ with probability $f(a|x,t)$ where

$$\sum_{a \in \mathbf{A}(x,t)} f(a|x,t) = 1 \ .$$

The set of all Markov strategies will be denoted by $\mathbf{B}$.

Theorem 3.9

Consider a finite-horizon Markov Decision Process and let s^* be a strategy found by dynamic programming. Then s^* is an optimal Markov strategy for that process.

Proof

The proof proceeds by induction on t (backward in time, naturally). Let $\pi_t(x|\beta)$ be the expected future payoff at time t for using a strategy β given that the decision process is in state x at that time. Assume that the backward induction strategy is optimal for some time $t + 1$:

$$\pi_{t+1}(x|\beta) \leq \pi_{t+1}(x|s^*) \equiv \pi_{t+1}^*(x) \qquad \forall x \in \mathbf{X} \text{ and } \forall \beta \in \mathbf{B}. \qquad (3.7)$$

3.4 Optimal Strategies for Finite Processes

Then (for arbitrary β)

$$\begin{aligned}
\pi_t(x|\beta) &= \sum_{a \in \mathbf{A}(x,t)} f(a|x,t) \left[r(x,a) + \sum_{x' \in \mathbf{X}} p(x'|x,a) \pi_{t+1}(x'|\beta) \right] \\
&\leq \sum_{a \in \mathbf{A}(x,t)} f(a|x,t) \left[r(x,a) + \sum_{x' \in \mathbf{X}} p(x'|x,a) \pi^*_{t+1}(x') \right] \\
&\leq \left(\sum_{a \in \mathbf{A}(x,t)} f(a|x,t) \right) \pi^*_t(x) \\
&= \pi^*_t(x)
\end{aligned}$$

where the first inequality follows from the inductive assumption (Equation 3.7) and the second follows from the stochastic dynamic programming equation. Because

$$\pi_T(x|\beta) = r_T(x) = \pi^*_T(x) \quad \forall x \in \mathbf{X}.$$

The inequality $\pi_t(x|\beta) \leq \pi^*_t(x)$ holds for all x and t. In particular, the optimality condition $\pi_0(x|\beta) \leq \pi^*_0(x) \, \forall x \in \mathbf{X}$ holds. □

Having established that the strategy found by dynamic programming is an optimal *Markov* strategy, we now consider whether enlarging the class of available strategies leads to a better strategy.

Definition 3.10

Denote the set of actions available in state x at time t by $\mathbf{A}(x,t)$. For each history $h_t(x)$ that leads to state x at time t, a *behavioural strategy* specifies using action $a \in \mathbf{A}(x,t)$ with probability $\phi(a|h_t(x),t)$, where

$$\sum_{a \in \mathbf{A}(x,t)} \phi(a|h_t(x),t) = 1.$$

Markov strategies are a subset of the set of behavioural strategies – the subset where decisions are conditioned only on the part of the history that specifies the current state.

Theorem 3.11

No behavioural strategy gives a higher payoff than the strategy found by dynamic programming.

Proof

The proof proceeds by showing that for every general behavioural strategy, there is a Markov strategy that gives the same expected payoff. Consequently, no behavioural strategy can do better than every Markov strategy. Because dynamic programming strategies are optimal Markov strategies, the desired result follows. See Filar & Vrieze (1997) for details. □

Exercise 3.4

Consider the MDP shown below with $T = 3$ and terminal rewards $r_3(x) = r_3(y) = r_3(z) = 0$. Find the optimal strategy.

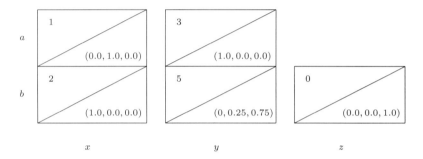

3.5 Infinite-horizon Markov Decision Processes

We now consider processes in which there is no a priori termination: the process continues forever, unless some decision is taken that has the consequence of ending the process.

When considering an infinite-horizon process, we might consider the payoff when starting in state x as being given by limit as $T \to \infty$ of the corresponding payoff for a finite-horizon process.

$$\pi(x|\sigma) = \mathbb{E}\left[\sum_{t=0}^{\infty} r_t(X_t, A_t)\right]$$

$$= \sum_{t=0}^{\infty} \mathbb{E}\left[r_t(X_t, A_t)\right].$$

Obviously, because there is no a priori termination of the process, there are no terminal rewards. The first question is: does the limit as $T \to \infty$ exist for all possible strategies?

Example 3.12

Consider a very simple process with only one state, x. In that state, the action set is $\mathbf{A} = \{a_1, a_2\}$ and $r(x, a_1) = 1$ and $r(x, a_2) = 2$, independent of time. Whichever action is chosen the process (necessarily) returns to x. Suppose we have two strategies: $s_1 =$ "always choose a_1" and $s_2 =$ "always choose a_2". Clearly, with the payoff defined as above, we have

$$\pi(x|s_1) = \pi(x|s_2) = \infty$$

so there is apparently no way of choosing the better strategy.

Instead of considering constant rewards, let us introduce the simple time dependence: $r_t(x, a) = \delta^t r(x, a)$. Provided the discount factor δ is such that $0 \leq \delta < 1$ the payoff

$$\pi(x|\sigma) = \mathbb{E} \sum_{t=0}^{\infty} \left[\delta^t r(X_t, A_t) \right] \tag{3.8}$$

is finite for all strategies σ and all initial states x.

Exercise 3.5

Consider the process described in Example 3.12. Find the payoffs for the strategies s_1 and s_2 if rewards are discounted by an amount $0 \leq \delta < 1$.

Why should discounting be introduced into a model? Apart from being a mathematical "trick" to ensure the finiteness of all payoffs as already discussed, there are two reasons. First, if the rewards are monetary, the discount factor δ models a (constant) depreciation of value due to inflation: one unit of currency next year will be worth less (i.e., buy less) than one unit of currency today. Second, the discount factor can be viewed as the probability that the decision process continues for at least one more time step. That is, with probability $1 - \delta$ some catastrophe occurs (independently of any strategy adopted) that terminates the decision process. In a biological context, $1 - \delta$ is the probability that the organism dies as a result of factors not being explicitly considered in the problem.

3.6 Optimal Strategies for Infinite Processes

The dynamic programming equation (including discount factor) for a finite horizon problem is

$$\pi_t^*(x) = \max_{a \in \mathbf{A}(x)} \left[r(x,a) + \delta \sum_{x' \in \mathbf{X}} p(x'|x,a) \pi_{t+1}^*(x') \right].$$

An infinite horizon model has the property that after a decision has been made an individual will find themselves facing the *same* infinite horizon decision problem as before, albeit starting in a different state. Therefore, it seems reasonable to guess that the infinite-horizon dynamic programming equation can be found by setting $\pi_t^*(x) = \pi_{t+1}^*(x) = \pi^*(x) \ \forall x \in \mathbf{X}$ in the equation above. The following theorem shows that this guess is correct.

Theorem 3.13

The optimal payoffs satisfy the dynamic programming equation

$$\pi^*(x) = \max_{a \in \mathbf{A}(x)} \left[r(x,a) + \delta \sum_{x' \in \mathbf{X}} p(x'|x,a) \pi^*(x') \right]. \tag{3.9}$$

Proof

Let σ be an arbitrary strategy that chooses action a at $t = 0$ with probability $f(a)$. Then

$$\pi(x|\sigma) = \sum_{a \in \mathbf{A}(x)} f(a) \left[r(x,a) + \delta \sum_{x' \in \mathbf{X}} p(x'|x,a) \pi_1(x'|\sigma) \right]$$

where $\pi_1(x'|\sigma)$ is the payoff that σ achieves starting from state x' at time $t = 1$. Because $\pi_1(x'|\sigma) \leq \pi^*(x')$ we have

$$\begin{aligned}
\pi(x|\sigma) &\leq \sum_{a \in \mathbf{A}(x)} f(a) \left[r(x,a) + \delta \sum_{x' \in \mathbf{X}} p(x'|x,a) \pi^*(x') \right] \\
&\leq \sum_{a \in \mathbf{A}(x)} f(a) \max_{a \in \mathbf{A}(x)} \left[r(x,a) + \delta \sum_{x' \in \mathbf{X}} p(x'|x,a) \pi^*(x') \right] \\
&= \max_{a \in \mathbf{A}(x)} \left[r(x,a) + \delta \sum_{x' \in \mathbf{X}} p(x'|x,a) \pi^*(x') \right].
\end{aligned}$$

3.6 Optimal Strategies for Infinite Processes

Because the inequality is true for arbitrary σ, it holds for the optimal strategy. Hence

$$\begin{aligned}\pi^*(x) &\equiv \pi(x|\sigma^*) \\ &\leq \max_{a \in \mathbf{A}(x)} \left[r(x,a) + \delta \sum_{x' \in \mathbf{X}} p(x'|x,a)\pi^*(x') \right].\end{aligned}$$

Now, in order to show that the opposite inequality also holds, let

$$\hat{a} \in \operatorname*{argmax}_{a \in \mathbf{A}(x)} \left[r(x,a) + \delta \sum_{x' \in \mathbf{X}} p(x'|x,a)\pi^*(x') \right]$$

and let $\sigma(\hat{a})$ be the strategy that chooses $\hat{a}$ at $t=0$ and then acts optimally for the process starting at time $t=1$. Then

$$\begin{aligned}\pi^*(x) &\geq \pi(x|\sigma(\hat{a})) \\ &= r(x,\hat{a}) + \delta \sum_{x' \in \mathbf{X}} p(x'|x,\hat{a})\pi^*(x') \\ &= \max_{a \in \mathbf{A}(x)} \left[r(x,a) + \delta \sum_{x' \in \mathbf{X}} p(x'|x,a)\pi^*(x') \right]\end{aligned}$$

Combining the two inequalities completes the proof. □

Now we show that the optimal payoff is the unique solution of Equation 3.9.

Theorem 3.14

The payoff $\pi^*(x)$ is the unique solution of Equation 3.9.

Proof

Suppose $\pi_1(x)$ and $\pi_2(x)$ both satisfy Equation 3.9. Then, setting

$$\hat{a}(x) \in \operatorname*{argmax}_{a \in \mathbf{A}(x)} \left[r(x,a) + \delta \sum_{x' \in \mathbf{X}} p(x'|x,a)\pi_1(x') \right]$$

we have for each state x

$$\begin{aligned}\pi_1(x) - \pi_2(x) &= \delta \max_{a \in \mathbf{A}(x)} \sum_{x' \in \mathbf{X}} p(x'|x,a) \left[\pi_1(x') - \pi_2(x') \right] \\ &= \delta \sum_{x' \in \mathbf{X}} p(x'|x,\hat{a}(x)) \left[\pi_1(x') - \pi_2(x') \right]\end{aligned}$$

$$\leq \delta \sum_{x' \in \mathbf{X}} p(x'|x, \hat{a}(x)) |\pi_1(x') - \pi_2(x')|$$

$$\leq \delta \sum_{x' \in \mathbf{X}} p(x'|x, \hat{a}(x)) \max_{x' \in \mathbf{X}} |\pi_1(x') - \pi_2(x')|$$

$$= \delta \max_{x' \in \mathbf{X}} |\pi_1(x') - \pi_2(x')| .$$

Let

$$x_m \in \operatorname*{argmax}_{x' \in \mathbf{X}} |\pi_1(x') - \pi_2(x')|$$

then we have

$$\pi_1(x_m) - \pi_2(x_m) \leq \delta |\pi_1(x_m) - \pi_2(x_m)| .$$

Reversing the roles of π_1 and π_2 gives

$$\pi_2(x_m) - \pi_1(x_m) \leq \delta |\pi_2(x_m) - \pi_1(x_m)| .$$

Combining the two inequalities yields

$$|\pi_1(x_m) - \pi_2(x_m)| \leq \delta |\pi_1(x_m) - \pi_2(x_m)|$$

and, because $\delta < 1$, we must have

$$|\pi_1(x_m) - \pi_2(x_m)| = \max_{x' \in \mathbf{X}} |\pi_1(x') - \pi_2(x')|$$
$$= 0 .$$

Hence

$$|\pi_1(x) - \pi_2(x)| = 0 \quad \forall x \in \mathbf{X} .$$

□

Now that we know that the optimal strategy gives the payoff which uniquely satisfies Equation 3.9 we can use this fact to prove that a stationary and non-randomising strategy is optimal.

Definition 3.15

Let s be a non-randomising and stationary strategy that selects action $a(x) \in \mathbf{A}(x)$ every time the process is in state x. Let $g(x) : \mathbf{X} \to \mathbb{R}$ be a bounded function (i.e., $g(x) < \infty \, \forall x \in \mathbf{X}$). Define an operator T_s by

$$(T_s g)(x) = r(x, a(x)) + \delta \sum_{x' \in \mathbf{X}} p(x'|x, a(x)) g(x').$$

Suppose we let T_s act on $g(x)$ and then let T_s act on the result of the first operation. We will denote the combined operation by $(T_s^2 g)(x)$. Similarly, the n-fold action of T_s will be denoted T_s^n.

3.6 Optimal Strategies for Infinite Processes

Lemma 3.16

For any bounded, real function $g(x)$, $\lim_{n\to\infty}(T_s^n g)(x) = \pi(x|s)$.

Proof

$$\begin{aligned}
(T_s^n g)(x) &= r(x,a(x)) + \delta \sum_{x'\in \mathbf{X}} p(x'|x,a(x))(T_s^{n-1}g)(x') \\
&= r(x,a(x)) + \delta \sum_{x'\in \mathbf{X}} p(x'|x,a(x))r(x',a(x')) \\
&\quad + \delta^2 \sum_{x''\in \mathbf{X}} p(x''|x',a(x'))(T_s^{n-2}g)(x'') \\
&= \mathbb{E}[r(X_0,a_0)|s] + \mathbb{E}[r(X_1,a_1)|s] \\
&\quad + \delta^2 \sum_{x''\in \mathbf{X}} p(x''|x',a(x'))(T_s^{n-2}g)(x'')
\end{aligned}$$

Continuing the expansion and because $\delta < 1$ we have

$$\begin{aligned}
\lim_{n\to\infty}(T_s^n g)(x) &= \sum_{t=0}^{\infty} \mathbb{E}[r(X_t,a_t)] \\
&\equiv \pi(x|s).
\end{aligned}$$

$\square$

Theorem 3.17

Let

$$a^*(x) \in \operatorname*{argmax}_{a\in \mathbf{A}(x)} \left[r(x,a) + \delta \sum_{x'\in \mathbf{X}} p(x'|x,a)\pi^*(x') \right] \quad \forall x \in \mathbf{X} \qquad (3.10)$$

and let s^* be the non-randomising and stationary strategy that selects $a^*(x)$ every time the process is in state x. Then s^* is optimal.

Proof

By the definition of s^* and using the dynamic programming equation we have

$$(T_{s^*}\pi^*)(x) = \pi^*(x)$$

which implies that

$$(T_{s^*}^n \pi^*)(x) = \pi^*(x) \quad \forall n.$$

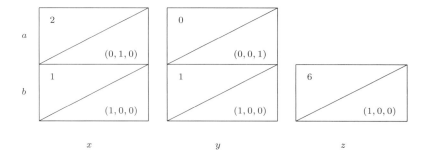

Figure 3.2 Diagrammatic representation of the Markov decision process for Exercise 3.6.

Now, letting $n \to \infty$ and using the result of Lemma 3.16, we have
$$\begin{aligned} \pi(x|s^*) &= \lim_{n \to \infty} (T_{s^*}^n \pi^*)(x) \\ &= \pi^*(x). \end{aligned}$$

$\square$

If we can guess the optimal strategy, then all we have to do is check that the actions $a(x)$ specified by that strategy satisfy Equation 3.10.

Exercise 3.6

Consider the MDP shown in Figure 3.2. Assuming that this as a discounted infinite-horizon problem with $\delta = \frac{1}{2}$, show that the optimal strategy is $a^*(x) = a$ and $a^*(y) = a$. (Because being in state z gives the highest reward, it seems worth trying a strategy that eventually puts the process in state z starting from any state.) [Hint: solve the dynamic programming equation to find the payoffs for following the specified strategy. Then show that changing the action chosen in any state gives a lower payoff.]

3.7 Policy Improvement

The dynamic programming equation suggests the following iterative procedure for finding an optimal strategy and its associated payoffs in an infinite-horizon MDP. This procedure is often called *Policy Improvement* because strategies are called "policies" by people who study MDPs but not games.

3.7 Policy Improvement

1. Start by picking an arbitrary strategy s that specifies using action $a(x)$ in state x.

2. Solve the set of simultaneous equations

$$\pi(x|s) = r(x,a) + \delta \sum_{x' \in \mathbf{X}} p(x'|x,a)\pi(x'|s)$$

   to find the payoffs for using that strategy.

3. Find an improved strategy by solving

$$\hat{a}(x) \in \operatorname*{argmax}_{a \in \mathbf{A}(x)} \left[r(x,a) + \delta \sum_{x' \in \mathbf{X}} p(x'|x,a)\pi(x'|s) \right].$$

4. Repeat steps 2 and 3 until the strategy doesn't change.

The following theorem proves that this algorithm works.

Theorem 3.18

Suppose we have some stationary pure strategy s that yields payoffs $\pi(x|s)$. Let

$$\hat{a}(x) \in \operatorname*{argmax}_{a \in \mathbf{A}(x)} \left[r(x,a) + \delta \sum_{x' \in \mathbf{X}} p(x'|x,a)\pi(x'|s) \right] \quad \forall x \in \mathbf{X}$$

and let $\hat{s}$ be the (non-randomising and stationary) strategy that selects $\hat{a}(x)$ every time the process is in state x. Then $\hat{s}$ is either a better strategy than s or both strategies are optimal.

Proof

Consider the operator $T_{\hat{s}}$ associated with the new strategy $\hat{s}$ acting on an arbitrary bounded function $g(x)$:

$$(T_{\hat{s}} g)(x) = r(x, \hat{a}(x)) + \delta \sum_{x' \in \mathbf{X}} p(x'|x, \hat{a}(x)) g(x').$$

From the definition of $\hat{a}(x)$, we have

$$T_{\hat{s}} \pi(x|s) \geq \pi(x|s) \quad \forall x \in \mathbf{X}.$$

Acting repeatedly on this inequality with $T_{\hat{s}}$ gives

$$T_{\hat{s}}^n \pi(x|s) \geq T_{\hat{s}}^{n-1} \pi(x|s) \geq \cdots \geq \pi(x|s).$$

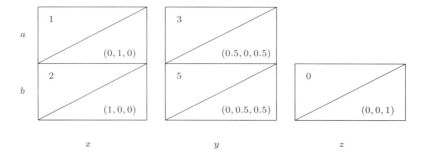

Figure 3.3 Diagrammatic representation of the Markov decision process solved in Example 3.19.

Now, letting $n \to \infty$ and using the result of Lemma 3.16, we have
$$\pi(x|\hat{s}) \geq \pi(x|s) \qquad \forall x \in \mathbf{X}.$$

So we have shown that the new strategy $\hat{s}$ is at least as good as the old one. Next we will show that if the strategy $\hat{s}$ is not strictly better than s in at least one state, then both strategies are optimal.

We have just established that
$$\pi(x|\hat{s}) \geq T_{\hat{s}}\pi(x|s) \geq \pi(x|s).$$

Now suppose that $\pi(x|\hat{s}) = \pi(x|s) \quad \forall x \in \mathbf{X}$. This implies that
$$\begin{aligned}\pi(x|\hat{s}) &= T_{\hat{s}}\pi(x|s) \\ &= T_{\hat{s}}\pi(x|\hat{s}) \\ &= \max_{a \in \mathbf{A}(x)} \left[r(x,a) + \delta \sum_{x' \in \mathbf{X}} p(x'|x,a)\pi(x'|\hat{s}) \right].\end{aligned}$$

So $\pi(x|\hat{s})$ satisfies the optimality equation. By the uniqueness of the solution to that equation (Theorem 3.14) we must have
$$\pi(x|\hat{s}) = \pi^*(x) = \pi(x|s)$$

which proves that s and $\hat{s}$ are both optimal strategies. □

Example 3.19

Consider the problem shown in Figure 3.2 with discount factor $\delta = \frac{9}{10}$. Let us begin with the strategy $s_0 = \{a(x) = b, a(y) = b\}$.

3.7 Policy Improvement

Iteration 1: The payoffs for s_0 are found from

$$\pi(x|s_0) = 2 + \delta\pi(x|s_0) \quad \text{and}$$
$$\pi(y|s_0) = 5 + \frac{1}{2}\delta\pi(y|s_0)$$

which give $\pi(x|s_0) = 20$ and $\pi(y|s_0) = \frac{100}{11}$.

Then the payoffs for changing the action taken in each state to a are

$$r(x,a) + \delta\pi(y|s_0) = 1 + \frac{9}{10}\frac{100}{11} = \frac{101}{11} \quad \text{and}$$
$$r(y,a) + \frac{1}{2}\delta\pi(x|s_0) = 3 + \frac{9}{20} \times 20 = 12$$

Because $\frac{101}{11} < 20$ and $12 > \frac{100}{11}$, we can conclude that $\hat{a}(x) = b$ and $\hat{a}(y) = a$. Let us call this new strategy s_1.

Iteration 2: The payoffs for s_1 are found from

$$\pi(x|s_1) = 2 + \delta\pi(x|s_1) \quad \text{and}$$
$$\pi(y|s_1) = 3 + \frac{1}{2}\delta\pi(x|s_1)$$

which give $\pi(x|s_1) = 20$ and $\pi(y|s_1) = 12$.

Then the payoffs for changing the action taken in each state are

$$r(x,a) + \delta\pi(y|s_1) = 1 + \frac{9}{10} \times 12 = \frac{118}{10} \quad \text{and}$$
$$r(y,b) + \frac{1}{2}\delta\pi(x|s_1) = 5 + \frac{9}{20} \times 12 = \frac{104}{10}$$

from which we can conclude that changing strategy does not yield a better payoff. Therefore

$$s^* = \{a(x) = b, a(y) = a\}$$

is an optimal strategy.

The optimal strategy that we have found is one that seems intuitively reasonable for $\delta \to 1$ because it reduces the probability that the process will end up in state z producing an "infinite" stream of zero rewards.

Exercise 3.7

Find the optimal strategy for the previous exercise by starting with $s = \{a(x) = a, a(y) = a\}$ or $s = \{a(x) = a, a(y) = b\}$.

Part II

Interaction

4
Static Games

4.1 Interactive Decision Problems

An interactive decision problem involves two or more individuals making a decision in a situation where the payoff to each individual depends (at least in principle) on what every individual decides. Borrowing some terminology from recreational games, which form only a subset of examples of interactive decision problems, all such problems are termed "games" and the individuals making the decisions are called "players". However, recreational games may have restrictive features that are not present in general games: for example, it is not necessarily true that one player "wins" only if the other "loses". Games that have winners and losers in this sense are called *zero-sum games*; these are considered in Section 4.7.3.

Definition 4.1

A *static game* is one in which a single decision is made by each player, and each player has no knowledge of the decision made by the other players before making their own decision.

Sometimes such games are referred to as *simultaneous decision games* because any actual order in which the decisions are made is irrelevant. The most famous example of an interactive decision problem is probably the Prisoners' Dilemma.

Example 4.2 (Prisoners' Dilemma)

Two crooks are being questioned by the police in connection with a serious crime. They are held in separate cells and cannot talk to each other. Without a confession, the police only have enough evidence to convict the two crooks on a lesser charge. The police make the following offer to both prisoners (in separate rooms so that no communication between them is possible): if one confesses that both committed the serious crime, then the confessor will be set free and the other will spend 5 years in jail (4 for the crime and 1 for obstructing justice); if both confess, then they will each get the 4-year sentence; if neither confess, then they will each spend 2 years in jail for the minor offense.

We can describe this game more succinctly using the following table of payoffs, where the possible courses of action open to each prisoner are (i) Q = "Keep Quiet" or (ii) S = "Squeal". The payoffs are given in terms of years of freedom lost. The payoffs for the first prisoner (P_1) are given first in each pair of entries in the table; those for the other prisoner (P_2) come second.

		P_2	
		Q	S
P_1	Q	−2, −2	−5, 0
	S	0, −5	−4, −4

What should each prisoner do? First, consider P_1. If P_2 keeps quiet, then they should squeal because that leads to 0 years in jail rather than 2 years. On the other hand, if P_2 squeals, then they should also squeal because that leads to 4 years in jail rather than 5. So whatever P_2 does, P_1 is better off if they squeal. Similarly, P_2 is better off squealing no matter what P_1 does. So both prisoners should squeal.

The interest in this game arises from the following observation. Both players, by following their individual self-interest, end up worse off than if they had kept quiet. This apparently paradoxical result encapsulates a major difference between non-interactive and interactive decision models (games). It might be argued that they should have had an agreement before being arrested that they wouldn't squeal ("honour among thieves"). However, each prisoner has no way of ensuring that the other follows this agreement. Of course, a prisoner could exact revenge in the future on a squealer – but that is another game (with different payoffs).

Definition 4.3

A solution is said to be *Pareto optimal* (after the Italian economist Vilfredo

Pareto) if no player's payoff can be increased without decreasing the payoff to another player. Such solutions are also termed *socially efficient* or just *efficient*.

The Prisoners' Dilemma is often used as the starting point for a discussion of the Social Contract (i.e., how societies form and how they are sustained) because the socially inefficient nature of its solution is reminiscent of many features of society. For example, consider paying taxes. Whatever anyone else does, you are better off (more wealthy) if you do not pay your taxes. However, if no-one pays any taxes (because, like you, they are following their own self-interest), then there is no money to provide community services and everyone is worse off than if everyone had paid their taxes.

Example 4.4 (Standardised Prisoners' Dilemma)

Any game of the form

$$
\begin{array}{c|c|c|}
 & \multicolumn{2}{c}{P_2} \\
 & C & D \\
\hline
P_1 \quad C & r,r & s,t \\
\hline
D & t,s & p,p \\
\hline
\end{array}
$$

with $t > r > p > s$ is called a Prisoners' Dilemma.[1] A particularly common version has payoffs given by $t = 5$, $r = 3$, $p = 1$, and $s = 0$. The available courses of action are generically called "cooperation" (C) and "defection" (D).[2] Analysis of this game tells us that we should expect both players to defect – a solution that is socially inefficient.

4.2 Describing Static Games

To describe a static game, you need to specify:

1. the set of players, indexed by $i \in \{1, 2, \ldots\}$;
2. a pure strategy set, $\mathbf{S}_i$, for each player;

---

[1] These letters are conventionally used to represent (t) the payoff for yielding to <u>t</u>emptation, (r) the <u>r</u>eward for cooperating, (p) the <u>p</u>unishment for defection, and (s) the payoff for being a <u>s</u>ucker and not retaliating to defection.
[2] In the original version, the two prisoners are playing a game against each other, not against the police. So keeping quiet can been viewed as cooperating and squealing can be seen as defecting.

3. payoffs for each player for every possible combination of pure strategies used by all players.

To keep the notation simple, we will concentrate on two-player games for most of this book. Games with more than two players are briefly considered in Section 4.8. In the two-player case, it is conventional to put the strategy of player 1 first so that the payoffs to player i are written

$$\pi_i(s_1, s_2) \quad \forall s_1 \in \mathbf{S}_1 \quad \text{and} \quad \forall s_2 \in \mathbf{S}_2.$$

Definition 4.5

A tabular description of a game, using pure strategies, is called the *normal form* or *strategic form* of a game.

Remark 4.6

It is important to note that the strategic form uses pure strategies to describe a game. For a static game, there is no real distinction between pure strategies and actions. However, the distinction will become important when we consider dynamic games. (See the discussion in Section 2.3 for the importance in non-interactive decision problems.)

Example 4.7

The strategic form of the Prisoners' Dilemma is the table shown in Example 4.2. The pure strategy sets are $\mathbf{S}_1 = \mathbf{S}_2 = \{Q, S\}$ and the payoffs are given in the table, e.g.,

$$\pi_1(Q,Q) = -2 \quad \pi_1(Q,S) = -5 \quad \pi_2(Q,S) = 0.$$

Definition 4.8

A *mixed strategy* for player i gives the probabilities that action $s \in \mathbf{S}_i$ will be played. A mixed strategy will be denoted σ_i and the set of all possible mixed strategies for player i will be denoted by $\mathbf{\Sigma}_i$.

Remark 4.9

If a player has a set of strategies $\mathbf{S} = \{s_a, s_b, s_c, \ldots\}$ then a mixed strategy can be represented as a vector of probabilities:

$$\sigma = (p(s_a), p(s_b), p(s_c), \ldots).$$

4.2 Describing Static Games

A pure strategy can then be represented as a vector where all the entries are zero except one. For example,

$$s_b = (0, 1, 0, \ldots) .$$

Mixed strategies can, therefore, be represented as linear combinations of pure strategies:

$$\sigma = \sum_{s \in \mathbf{S}} p(s) s .$$

Usually, we will denote the probability of using pure strategy s by $p(s)$ for player 1 and and $q(s)$ for player 2. The payoffs for mixed strategies are then given by

$$\pi_i(\sigma_1, \sigma_2) = \sum_{s_1 \in \mathbf{S}_1} \sum_{s_2 \in \mathbf{S}_2} p(s_1) q(s_2) \pi_i(s_1, s_2) .$$

As usual, the payoffs are assumed to be a representation of the preferences of rational individuals or of their biological fitness, so that an individual's aim is to maximise their payoff (see Sections 1.3 and 1.4). As we have already seen in Example 4.2, this "maximisation" has to take into account the behaviour of the other player and, as a result, the payoff achieved by any player may not be the maximum of the available payoffs.

Notation 4.10

A solution of a game is a (not necessarily unique) pair of strategies that a rational pair of players might use. Solutions will be denoted by enclosing a strategy pair within brackets, such as (A, B) or (σ_1, σ_2), where we will put the strategy adopted by player 1 first. For example, the solution of the Prisoners' Dilemma can be represented by (S, S).

Exercise 4.1

In Puccini's opera *Tosca*, Tosca's lover has been condemned to death. The police chief, Scarpia, offers to fake the execution if Tosca will sleep with him. The bargain is struck. However, in order to keep her honour, Tosca stabs and kills Scarpia. Unfortunately, Scarpia has also reneged on the deal and Tosca's lover has been executed. Construct a game theoretic representation of this operatic plot.

4.3 Solving Games Using Dominance

Because we solved the Prisoners' Dilemma in an intuitively simple manner by observing that the strategy of "Squealing" was always better than "Keeping Quiet", it seems reasonable to attempt to solve games by eliminating poor strategies for each player.

Definition 4.11

A strategy for player 1, σ_1, is *strictly dominated* by σ_1' if

$$\pi_1(\sigma_1', \sigma_2) > \pi_1(\sigma_1, \sigma_2) \quad \forall \sigma_2 \in \Sigma_2 .$$

That is, whatever player 2 does, player 1 is always better off using σ_1' rather than σ_1. Similarly, a strategy for player 2, σ_2, is *strictly dominated* by σ_2' if

$$\pi_2(\sigma_1, \sigma_2') > \pi_2(\sigma_1, \sigma_2) \quad \forall \sigma_1 \in \Sigma_1 .$$

Definition 4.12

A strategy for player 1, σ_1, is *weakly dominated* by σ_1' if

$$\pi_1(\sigma_1', \sigma_2) \geq \pi_1(\sigma_1, \sigma_2) \quad \forall \sigma_2 \in \Sigma_2$$

and

$$\exists \sigma_2' \in \Sigma_2 \text{ s.t. } \pi_1(\sigma_1', \sigma_2') > \pi_1(\sigma_1, \sigma_2') .$$

A similar definition applies for player 2.

We have already solved the Prisoners' Dilemma by the elimination of strictly dominated strategies. The following example illustrates the solution of a game by the elimination of weakly dominated strategies.

Example 4.13

Consider the following game.

		P_2	
		L	R
P_1	U	3, 3	2, 2
	D	2, 1	2, 1

4.3 Solving Games Using Dominance

For player 1, U weakly dominates D and, for player 2, L weakly dominates R. Consequently, we expect that player 1 will not play D and player 2 will not play R, leaving the solution (U, L).

To solve a game by the elimination of dominated strategies we have to assume that the players are rational. However, we can go further, if we also assume that:

1. The players are rational.
2. The players all know that the other players are rational.
3. The players all know that the other players know that they are rational.
4. ...(in principle) *ad infinitum*.

This chain of assumptions is called *Common Knowledge of Rationality*, or CKR. It encapsulates the idea of being able to "put oneself in another's shoes". By applying the CKR assumption, we can solve a game by iterating the elimination of dominated strategies.

Example 4.14

Consider the following game:

		P_2		
		L	M	R
P_1	U	1,0	1,2	0,1
	D	0,3	0,1	2,0

Initially player 1 has no dominated strategies. For player 2, R is dominated by M. So R is eliminated as a reasonable strategy for player 2. Now, for player 1, D is dominated by U. So D is eliminated as a reasonable strategy for player 1. Now, for player 2, L is dominated by M. Eliminating L, leaves (U, M) as the unique solution. (The levels of CKR listed explicitly above have been used in this example.)

There is a problem with the iterated elimination of dominated strategies when it comes to dealing with weakly dominated strategies: the solution may depend on the order in which strategies are eliminated.

Example 4.15

Consider the following game:

	P_2		
	L	M	R
U	10, 0	5, 1	4, −2
D	10, 1	5, 0	1, −1

(with P_1 labeling the rows)

Order 1: Eliminate D for player 1. Now eliminate L and R for player 2. The remaining strategy pair (U, M) is postulated as the solution, but using a different order of elimination we arrive at a different result. *Order 2:* Eliminate R for player 2. Neither player now has any dominated strategies, so stop. There are four remaining strategy pairs which could be the solution to the game, namely (U, L), (U, M), (D, L) and (D, M).

Exercise 4.2

Solve the following abstract games using the (iterated) elimination of dominated strategies. For the second game, does the solution depend on the order of elimination?

(a)

	P_2	
	L	R
U	3, 0	2, 1
D	2, 1	1, 0

(b)

	P_2	
	L	R
U	0, 3	10, 2
C	10, 4	0, 0
D	3, 1	3, 1

4.4 Nash Equilibria

The next example shows that some games can only be trivially solved using the (iterated) elimination of dominated strategies.

Example 4.16

Consider the game:

	P_2		
	L	M	R
U	1, 3	4, 2	2, 2
C	4, 0	0, 3	4, 1
D	2, 5	3, 4	5, 6

From the start, neither player has any dominated strategies leading to the

4.4 Nash Equilibria

maximally imprecise prediction that anything can happen. (It is in this sense that the solution is "trivial".)

Nevertheless, there is an "obvious" solution to this game, namely (D, R), which maximises the payoff to both players. Is it possible to define a solution in terms of something other than the (iterated) elimination of dominated strategies that both identifies such obvious solutions and keeps many of the results derived using dominance techniques? Fortunately, the answer to this question is "yes": such a solution can be provided by the definition of a Nash equilibrium.

Definition 4.17

A *Nash equilibrium* (for two player games) is a pair of strategies (σ_1^*, σ_2^*) such that
$$\pi_1(\sigma_1^*, \sigma_2^*) \geq \pi_1(\sigma_1, \sigma_2^*) \qquad \forall \sigma_1 \in \Sigma_1$$
and
$$\pi_2(\sigma_1^*, \sigma_2^*) \geq \pi_2(\sigma_1^*, \sigma_2) \qquad \forall \sigma_2 \in \Sigma_2 .$$

In other words, given the strategy adopted by the other player, neither player could do strictly better (i.e., increase their payoff) by adopting another strategy.

Example 4.18

Consider the game from Example 4.16. Let $\sigma_2^* = R$ and let $\sigma_1 = (p, q, 1-p-q)$ (that is, σ_1 is an arbitrary strategy that specifies using U with probability p, C with probability q and D with probability $1 - p - q$). Then
$$\begin{aligned} \pi_1(\sigma_1, R) &= 2p + 4q + 5(1 - p - q) \\ &= 5 - 3p - q \\ &\leq 5 \\ &= \pi_1(D, R) . \end{aligned}$$

Now let $\sigma_1^* = D$ and let $\sigma_2 = (p, q, 1 - p - q)$. Then
$$\begin{aligned} \pi_2(D, \sigma_2) &= 5p + 4q + 6(1 - p - q) \\ &= 6 - p - 2q \\ &\leq 6 \\ &= \pi_2(D, R) . \end{aligned}$$

Consequently the pair (D, R) constitutes a Nash equilibrium.

Exercise 4.3

Consider the following game. Show that (D, L) and (U, M) are Nash equilibria.

$$P_1 \quad \begin{array}{c|c|c|c} & \multicolumn{3}{c}{P_2} \\ & L & M & R \\ \hline U & 10, 0 & 5, 1 & 4, -2 \\ \hline D & 10, 1 & 5, 0 & 1, -1 \end{array}$$

It is clear from Definition 4.17 and the previous exercise that a Nash equilibrium never includes strictly dominated strategies, but it *may* include weakly dominated strategies.

An alternative form of the definition of a Nash equilibrium is useful for finding Nash equilibria rather than just checking that a particular pair of strategies is a Nash equilibrium. First we define the concept of a best response strategy.

Definition 4.19

A strategy for player 1, $\hat{\sigma}_1$, is a *best response* to some (fixed) strategy for player 2, σ_2, if

$$\hat{\sigma}_1 \in \operatorname*{argmax}_{\sigma_1 \in \Sigma_1} \pi_1(\sigma_1, \sigma_2) .$$

Similarly, $\hat{\sigma}_2$ is a best response to some σ_1 if

$$\hat{\sigma}_2 \in \operatorname*{argmax}_{\sigma_2 \in \Sigma_2} \pi_2(\sigma_1, \sigma_2) .$$

An equivalent form of the definition of a Nash equilibrium, which focusses on the strategies rather than the payoffs, is that σ_1^* is a best response to σ_2^* and vice versa.

Definition 4.20

A pair of strategies (σ_1^*, σ_2^*) is a Nash equilibrium if

$$\sigma_1^* \in \operatorname*{argmax}_{\sigma_1 \in \Sigma_1} \pi_1(\sigma_1, \sigma_2^*)$$

and

$$\sigma_2^* \in \operatorname*{argmax}_{\sigma_2 \in \Sigma_2} \pi_2(\sigma_1^*, \sigma_2) .$$

It is clear that a strictly dominated strategy is never a best response to any strategy, whereas a weakly dominated strategy may be a best response to

4.4 Nash Equilibria

some strategy. This is why weakly dominated strategies may appear in Nash equilibria but strictly dominated strategies do not.

To use this definition to find Nash equilibria we find, for each player, the set of best responses to every possible strategy of the other player. We then look for pairs of strategies that are best responses to each other.

Example 4.21 (Matching Pennies)

Two players each place a penny[3] on a table, either "heads up" (strategy H) or "tails up" (strategy T). If the pennies match, player 1 wins (the pennies); if the pennies differ, then player 2 wins (the pennies).

		P_2	
		H	T
P_1	H	$+1, -1$	$-1, +1$
	T	$-1, +1$	$+1, -1$

Clearly, this is a game in which the two players have completely opposing interests: one player only wins a penny when the other loses a penny. Because a penny is a small amount of money (and anyway the coins may be used only as a token for playing, with each player retaining their own coin), the payoff may be interpreted as a utility (based on the pleasure of winning) of $+1$ for winning the game and a utility of -1 for losing.

We can easily check that there is no pure strategy pair that is a Nash equilibrium: (H, H) is not an equilibrium because P_2 should switch to T; (H, T) is not an equilibrium because P_1 should switch to T; (T, H) is not an equilibrium because P_1 should switch to H; and, finally, (T, T) is not an equilibrium because P_2 should switch to H. (Intuitively, the solution is obvious: each player should randomise – by tossing the penny – and play H with probability $\frac{1}{2}$.)

Let us consider the mixed strategies $\sigma_1 = (p, 1-p)$ for player 1 and $\sigma_2 = (q, 1-q)$. That is, player 1 plays "Heads" with probability p and player 2 plays "Heads" with probability q. The payoff to player 1 is

$$\pi_1(\sigma_1, \sigma_2) = pq - p(1-q) - (1-p)q + (1-p)(1-q)$$
$$= 1 - 2q + 2p(2q - 1)$$

Clearly, if $q < \frac{1}{2}$ then player 1's best response is to choose $p = 0$ (i.e., $\hat{\sigma}_1 = (0, 1)$ or "play Tails"). On the other hand, if $q > \frac{1}{2}$ then player 1's best response is to choose $p = 1$ (i.e., $\hat{\sigma}_1 = (1, 0)$ or "play Heads"). If $q = \frac{1}{2}$ then *every* mixed (and pure) strategy is a best response.

---

[3] In parts of Europe, you could use cents.

Now consider the payoff to player 2.

$$\begin{aligned}\pi_2(\sigma_1, \sigma_2) &= -pq + p(1-q) + (1-p)q - (1-p)(1-q) \\ &= -1 + 2p + 2q(1-2p)\end{aligned}$$

Clearly, if $p < \frac{1}{2}$ then player 2's best response is to choose $q = 1$ (i.e., $\hat{\sigma}_2 = (1,0)$ or "play Heads"). On the other hand, if $p > \frac{1}{2}$ then player 2's best response is to choose $q = 0$ (i.e., $\hat{\sigma}_2 = (0,1)$ or "play Tails"). If $p = \frac{1}{2}$ then *every* mixed (and pure) strategy is a best response.

So the only pair of strategies for which each is best response to the other is $\sigma_1^* = \sigma_2^* = (\frac{1}{2}, \frac{1}{2})$. That is,

$$[\sigma_1^*, \sigma_2^*] = \left[\left(\frac{1}{2}, \frac{1}{2}\right), \left(\frac{1}{2}, \frac{1}{2}\right)\right]$$

is a Nash equilibrium and the expected payoffs for each player are

$$\pi_1(\sigma_1^*, \sigma_2^*) = \pi_2(\sigma_1^*, \sigma_2^*) = 0 .$$

Remark 4.22

In contrast to single-player decision models (see Theorem 1.32), there is no solution to the Matching Pennies game involving only non-randomising strategies. In any given realisation of the Matching Pennies game, the outcome will be one of (H,H), (H,T), (T,H), or (T,T) each with probability $\frac{1}{4}$. The outcome of a game occurs as a result of the strategies chosen by the players, but a player's strategy is not the same as a choice of outcome.

Exercise 4.4

Find all the Nash equilibria of the following games.

(a)

		P_2	
		L	R
P_1	U	4, 3	2, 2
	D	2, 2	1, 1

(b)

		P_2	
		R	W
P_1	F	0, 0	2, 1
	M	1, 2	0, 0

We can often simplify the process of finding Nash equilibria by making use of the next two theorems. The first of these theorems makes it easy to find pure-strategy Nash equilibria.

Theorem 4.23

Suppose there exists a pair of pure strategies (s_1^*, s_2^*) such that

$$\pi_1(s_1^*, s_2^*) \geq \pi_1(s_1, s_2^*) \quad \forall s_1 \in \mathbf{S}_1$$

4.4 Nash Equilibria

and $\pi_2(s_1^*, s_2^*) \geq \pi_2(s_1^*, s_2) \quad \forall s_2 \in \mathbf{S}_2$.

Then (s_1^*, s_2^*) is a Nash equilibrium.

Proof

For all $\sigma_1 \in \Sigma_1$ we have

$$\begin{aligned}
\pi_1(\sigma_1, s_2^*) &= \sum_{s \in \mathbf{S}_1} p(s) \pi_1(s_1, s_2^*) \\
&\leq \sum_{s \in \mathbf{S}_1} p(s) \pi_1(s_1^*, s_2^*) \\
&= \pi_1(s_1^*, s_2^*) \, .
\end{aligned}$$

For all $\sigma_2 \in \Sigma_2$ we have

$$\begin{aligned}
\pi_2(s_1^*, \sigma_2) &= \sum_{s \in \mathbf{S}_2} q(s) \pi_2(s_1^*, s_2) \\
&\leq \sum_{s \in \mathbf{S}_2} q(s) \pi_1(s_1^*, s_2^*) \\
&= \pi_2(s_1^*, s_2^*) \, .
\end{aligned}$$

Hence (s_1^*, s_2^*) is a Nash equilibrium. $\qquad\square$

Example 4.24

Consider again the game from Example 4.16

$$P_1 \quad \begin{array}{c|c|c|c} & \multicolumn{3}{c}{P_2} \\ \hline & L & M & R \\ \hline U & 1,\underline{3} & \underline{4},2 & 2,2 \\ \hline C & \underline{4},0 & 0,\underline{3} & 4,1 \\ \hline D & 2,5 & 3,4 & \underline{5},\underline{6} \end{array}$$

Payoffs corresponding to a pure strategy that is a best response to one of the opponent's pure strategies are underlined. Two underlinings coincide in the entry $(5,6)$ corresponding to the strategy pair (D, R). The coincidence of underlinings means that D is a best response to R and vice versa (i.e., the pair of pure strategies (D, R) is a Nash equilibrium).

Exercise 4.5

Find the pure strategy Nash equilibria for the following game.

		P_2		
		L	M	R
P_1	U	4, 3	2, 7	0, 4
	D	5, 5	5, −1	−4, −2

Exercise 4.6

A man has two sons. When he dies, the value of his estate (after tax) is £1000. In his will it states that the two sons must each specify a sum of money s_i that they are willing to accept. If $s_1 + s_2 \leq 1000$, then each gets the sum he asked for and the remainder (if there is any) goes to the local home for spoilt cats. If $s_1 + s_2 > 1000$, then neither son receives any money and the entire sum of £1000 goes to the cats' home. Assume that (i) the two men care only about the amount of money they will inherit, and (ii) they can only ask for whole pounds. Find all the pure strategy Nash equilibria of this game.

In the process of finding the Nash equilibrium in the Matching Pennies game (see Example 4.21), we saw that, for each player, *any* strategy was a best response to the Nash equilibrium strategy of the other player. In particular, the payoff for playing H is equal to the payoff for playing T. Intuitively, the reason for this is obvious: if the payoffs were not equal, then player i could do better than the supposed mixed Nash equilibrium strategy σ_i^* by playing the pure strategy that assigns probability 1 to whichever of H or T gives the higher payoff. The following theorem shows that this result is generally true for all two-player games.

Definition 4.25

The *support* of a strategy σ is the set $\mathbf{S}(\sigma) \subseteq \mathbf{S}$ of all the strategies for which σ specifies $p(s) > 0$.

Example 4.26

Suppose an individual's pure strategy set is $\mathbf{S} = \{L, M, R\}$. Consider a mixed strategy of the form $\sigma = (p, 1-p, 0)$ where the probabilities are listed in the same order as the set $\mathbf{S}$ and $0 < p < 1$. The support of σ is $\mathbf{S}(\sigma) = \{L, M\}$.

Theorem 4.27 (Equality of Payoffs)

Let (σ_1^*, σ_2^*) be a Nash equilibrium, and let $\mathbf{S}_1^*$ be the support of σ_1^*. Then $\pi_1(s, \sigma_2^*) = \pi_1(\sigma_1^*, \sigma_2^*) \quad \forall s \in \mathbf{S}_1^*$.

4.4 Nash Equilibria

Proof

If the set $\mathbf{S}_1^*$ contains only one strategy, then the theorem is trivially true. Suppose now that the set $\mathbf{S}_1^*$ contains more than one strategy. If the theorem is not true, then at least one strategy gives a higher payoff to player 1 than $\pi_1(\sigma_1^*, \sigma_2^*)$. Let s' be the action that gives the greatest such payoff. Then

$$\begin{aligned}
\pi_1(\sigma_1^*, \sigma_2^*) &= \sum_{s \in \mathbf{S}_1^*} p^*(s) \pi_1(s, \sigma_2^*) \\
&= \sum_{s \neq s'} p^*(s) \pi_1(s, \sigma_2^*) + p^*(s') \pi_1(s', \sigma_2^*) \\
&< \sum_{s \neq s'} p^*(s) \pi_1(s', \sigma_2^*) + p^*(s') \pi_1(s', \sigma_2^*) \\
&= \pi_1(s', \sigma_2^*)
\end{aligned}$$

which contradicts the original assumption that (σ_1^*, σ_2^*) is a Nash equilibrium. $\square$

The corresponding result for player 2 also holds. Namely, if σ_2^* has support $\mathbf{S}_2^*$, then

$$\pi_2(\sigma_1^*, s) = \pi_2(\sigma_1^*, \sigma_2^*) \qquad \forall s \in \mathbf{S}_2^*.$$

The proof is analogous.

Remark 4.28

Because all strategies $s \in \mathbf{S}_1^*$ give the same payoff as the randomising strategy σ_1^*, why does player 1 (or indeed player 2) randomise? The answer is that, if player 1 were to deviate from this strategy, then σ_2^* would no longer be a best response and the equilibrium would disintegrate. This is why randomising strategies are important for games, in a way that they weren't for the single-player optimisation problems covered in Part I.

We can use Theorem 4.27 to find mixed strategy Nash equilibria.

Example 4.29

Consider the Matching Pennies game in Example 4.21. Suppose player 2 plays H with probability q and T with probability $1 - q$. If player 1 is playing a completely mixed strategy at the Nash equilibrium, then

$$\begin{aligned}
\pi_1(H, \sigma_2^*) &= \pi_1(T, \sigma_2^*) \\
\iff q\pi_1(H, H) + (1-q)\pi_1(H, T) &= q\pi_1(T, H) + (1-q)\pi_1(T, T)
\end{aligned}$$

$$\begin{aligned}
\iff q - (1-q) &= -q + (1-q) \\
\iff 4q &= 2 \\
\iff q &= \frac{1}{2}.
\end{aligned}$$

The same argument applies with the players swapped over, so the Nash equilibrium is (σ_1^*, σ_2^*) with $\sigma_1^* = \sigma_2^* = (\frac{1}{2}, \frac{1}{2})$ as we found before.

Exercise 4.7

Consider the children's game "Rock-Scissors-Paper", where 2 children simultaneously make a hand sign corresponding to one of the three items. Playing "Rock" (R) beats "Scissors" (S), "Scissors" beats "Paper" (P), and "Paper" beats "Rock". When both children play the same action (both R, both S, or both P) the game is drawn. (a) Construct a payoff table for this game with a payoff of $+1$ for a win, -1 for losing, and 0 for a draw. (b) Solve this game.

4.5 Existence of Nash Equilibria

John Forbes Nash Jr. proved the following theorem in 1950 as part of his PhD thesis, which is why equilibrium solutions to games are called "Nash equilibria".

Theorem 4.30 (Nash's Theorem)

Every game that has a finite strategic form (i.e., with finite number of players and finite number of pure strategies for each player) has at least one Nash equilibrium (involving pure or mixed strategies).

Remark 4.31

A general proof of Nash's theorem relies on the use of a fixed point theorem (e.g., Brouwer's or Kakutani's). Roughly, these fixed point theorems state that for some compact set $\mathbf{S}$ and a map $f \colon \mathbf{S} \to \mathbf{S}$ that satisfies various conditions, the map has a fixed point, i.e., that $f(p) = p$ for some $p \in \mathbf{S}$. The proof of Nash's theorem then amounts to showing that the best response map satisfies the necessary conditions for it to have a fixed point. Rather than spending a great deal of effort to prove one of the fixed point theorems, it seems preferable to restrict our attention to a class of games that is common and for which it is easy to provide a self-contained proof. We refer the interested reader to the

4.5 Existence of Nash Equilibria

more general proofs contained in the books by Fudenberg & Tirole (1993) and Myerson (1991).

Proposition 4.32

Every two player, two action game has at least one Nash equilibrium.

Proof

Consider a two player, two action game with arbitrary payoffs:

$$
\begin{array}{c|c|c|}
 & \multicolumn{2}{c}{P_2} \\
 & L & R \\
\hline
P_1 \quad U & a,b & c,d \\
\hline
D & e,f & g,h \\
\hline
\end{array}
$$

First we consider pure-strategy Nash equilibria: if $a \geq e$ and $b \geq d$ then (U, L) is a Nash equilibrium; if $e \geq a$ and $f \geq h$ then (D, L) is a Nash equilibrium; if $c \geq g$ and $d \geq b$ then (U, R) is a Nash equilibrium; if $g \geq c$ and $h \geq f$ then (D, R) is a Nash equilibrium. There is no pure strategy Nash equilibrium if either

1. $a < e$ and $f < h$ and $g < c$ and $d < b$, or
2. $a > e$ and $f > h$ and $g > c$ and $d > b$.

In these cases, we look for a mixed strategy Nash equilibrium using the Equality of Payoffs theorem (Theorem 4.27). Let $\sigma_1^* = (p^*, 1-p^*)$ and $\sigma_2^* = (q^*, 1-q^*)$. Then

$$
\begin{aligned}
\pi_1(U, \sigma_2^*) &= \pi_1(D, \sigma_2^*) \\
\Longleftrightarrow aq^* + c(1-q^*) &= eq^* + g(1-q^*) \\
\Longleftrightarrow q^* &= \frac{(c-g)}{(c-g)+(e-a)}
\end{aligned}
$$

and

$$
\begin{aligned}
\pi_2(\sigma_1^*, L) &= \pi_2(\sigma_1^*, R) \\
\Longleftrightarrow bp^* + f(1-p^*) &= dp^* + h(1-p^*) \\
\Longleftrightarrow p^* &= \frac{(h-f)}{(h-f)+(b-d)}
\end{aligned}
$$

In both cases, we have $0 < p^*, q^* < 1$ as required for a mixed strategy Nash equilibrium. □

Exercise 4.8

A general symmetric, 2 player, two strategy game has a payoff table

$$
\begin{array}{c|c|c|}
 & \multicolumn{2}{c}{P_2} \\
 & A & B \\
\hline
A & a, a & b, c \\
\hline
B & c, b & d, d \\
\hline
\end{array}
$$

P_1 on the left.

Show that such a game always has at least one symmetric Nash equilibrium.

4.6 The Problem of Multiple Equilibria

Some games have multiple Nash equilibria and, therefore, more than one possible solution.

Example 4.33 (Battle of the Sexes)

This is the classic example of a *coordination game*.[4] One modern version of the story is that a married couple are trying to decide what to watch on television. The husband would like to watch the football match and the wife would like to watch the soap opera. The total values of their utilities are made up of two increments. If they watch the programme of their choice, they get a utility increment of 1 (and zero otherwise). If they watch television together, each gets a utility increment of 2, whereas they get zero if they watch television apart — obviously they must be a rich couple with two TVs. So, using the pure strategy set $S =$ "watch soap opera" and $F =$ "watch football", the payoff table is

Husband / Wife:

	F	S
F	3,2	1,1
S	0,0	2,3

Clearly, this game has two pure-strategy Nash equilibria: (F, F) and (S, S).

---
[4] For biologists, the "Battle of the Sexes" is a different game — one that has no pure-strategy Nash equilibria. See Maynard Smith (1982).

4.6 The Problem of Multiple Equilibria

There is also a mixed strategy Nash equilibrium (σ_h^*, σ_w^*) with

$$\sigma_h^* = (p(F), p(S)) = \left(\frac{3}{4}, \frac{1}{4}\right)$$

$$\sigma_w^* = (q(F), q(S)) = \left(\frac{1}{4}, \frac{3}{4}\right).$$

This mixed strategy Nash equilibrium can be found using the Equality of Payoffs theorem (Theorem 4.27) or the best response method of Section 4.4 (which also finds the two pure-strategy equilibria).

In this game there is a problem with deciding what strategies will be adopted by the players. How should the players decide between these three Nash equilibria? Can they both decide on the same one? (This is not a problem with Game Theory itself: it just demonstrates that even simple interactive decision problems do not necessarily have simple solutions.) Note that for the randomising Nash equilibrium, the asymmetric outcomes can occur. The most likely outcome of the game if both players randomise is (F, S), which occurs with probability $\frac{9}{16}$, despite the fact that both players would prefer to coordinate.

Responses to the existence of multiple Nash equilibria have included:

1. Using a convention. For example, in the Battle of the Sexes, possible conventions are

    a) The man will get what he wants, because women are generous.

    b) The man should defer to the woman, because that's what a gentleman should do.

    c) ...

    This then leads to the question of which convention will be used and to the development of game-theoretic models of convention formation.

2. Refine the definition of a Nash equilibrium to eliminate some of the equilibria from consideration. There have been several attempts to do this ("trembling hand perfection", etc.) but, despite the inherent interest of such refinements, they do not succeed in eliminating all but one equilibrium in every case either one, many or (unfortunately) *no* refined equilibria may exist.

3. Invoke the concept of evolution: there is a population of players who pair up at various points in time to play this game. The proportion of players using any given strategy changes over time depending on the success of that strategy (either successful strategies are consciously imitated, or Natural Selection sorts it out). The evolutionary interpretation of Nash equilibria

can be viewed as a refinement of the Nash equilibrium concept (because it favours some equilibria over others). However, it is also important in its own right because of the application of game theory to evolutionary biology.

4.7 Classification of Games

4.7.1 Affine Transformations

If it is only the equilibrium strategies, and not the payoffs, which are of interest, then it is possible to convert a difficult calculation into a simpler one by means of a generalised affine transformation.

Definition 4.34

A *generalised affine transformation* of the payoffs for player 1 is

$$\pi'_1(s_1, s_2) = \alpha_1 \pi_1(s_1, s_2) + \beta_1(s_2) \qquad \forall s_1 \in \mathbf{S}_1$$

where $\alpha_1 > 0$ and $\beta_1(s_2) \in \mathbb{R}$.[5] Note that we may apply a different transformation for each possible pure strategy of player 2. Similarly, an affine transformation of the payoffs for player 2 is

$$\pi'_2(s_1, s_2) = \alpha_2 \pi_2(s_1, s_2) + \beta_2(s_1) \qquad \forall s_2 \in \mathbf{S}_2 .$$

Example 4.35

The game

		P_2	
		L	R
P_1	U	3,3	0,0
	D	−1,2	2,8

can be transformed into

		P_2	
		L	R
P_1	U	2,1	0,0
	D	0,0	1,2

---
[5] A standard affine transformation has $\beta_i(\cdot) = $ constant.

4.7 Classification of Games

by applying the affine transformations

$$\alpha_1 = \frac{1}{2} \qquad \beta_1(L) = \tfrac{1}{2} \qquad \beta_1(R) = 0$$
$$\alpha_2 = \frac{1}{3} \qquad \beta_2(U) = 0 \qquad \beta_2(D) = -\frac{2}{3}.$$

Exercise 4.9

Demonstrate by explicit calculation that the two games in Example 4.35 have the same Nash equilibria.

Theorem 4.36

If the payoff table is altered by generalised affine transformations, the set of Nash equilibria is unaffected.[6]

Proof

For player 1, we have

$$\pi'_1(\sigma_1^*, \sigma_2^*) \geq \pi'_1(\sigma_1, \sigma_2^*)$$
$$\iff \sum_{s_1}\sum_{s_2} p^*(s_1)q^*(s_2)\pi'_1(s_1, s_2) \geq \sum_{s_1}\sum_{s_2} p(s_1)q^*(s_2)\pi'_1(s_1, s_2)$$
$$\iff \alpha_1 \sum_{s_1}\sum_{s_2} p^*(s_1)q^*(s_2)\pi_1(s_1, s_2) + \sum_{s_1}\sum_{s_2} p^*(s_1)q^*(s_2)\beta(s_2)$$
$$\geq \alpha_1 \sum_{s_1}\sum_{s_2} p(s_1)q^*(s_2)\pi_1(s_1, s_2) + \sum_{s_1}\sum_{s_2} p(s_1)q^*(s_2)\beta(s_2)$$
$$\iff \alpha_1 \sum_{s_1}\sum_{s_2} p^*(s_1)q^*(s_2)\pi_1(s_1, s_2) + \sum_{s_2} q^*(s_2)\beta(s_2)$$
$$\geq \alpha_1 \sum_{s_1}\sum_{s_2} p(s_1)q^*(s_2)\pi_1(s_1, s_2) + \sum_{s_2} q^*(s_2)\beta(s_2)$$
$$\iff \alpha_1 \sum_{s_1}\sum_{s_2} p^*(s_1)q^*(s_2)\pi_1(s_1, s_2) \geq \alpha_1 \sum_{s_1}\sum_{s_2} p(s_1)q^*(s_2)\pi_1(s_1, s_2)$$
$$\iff \sum_{s_1}\sum_{s_2} p^*(s_1)q^*(s_2)\pi_1(s_1, s_2) \geq \sum_{s_1}\sum_{s_2} p(s_1)q^*(s_2)\pi_1(s_1, s_2)$$
$$\iff \pi_1(\sigma_1^*, \sigma_2^*) \geq \pi_1(\sigma_1, \sigma_2^*).$$

The analogous argument for player 2 completes the proof. □

---
[6] Remember that the payoffs at those equilibria *do* change.

4.7.2 Generic and Non-generic Games

Definition 4.37

A *generic game* is one in which a small change to any *one* of the payoffs[7] does not introduce new Nash equilibria or remove existing ones. In practice, this means that there should be no equalities between the payoffs that are compared to determine a Nash equilibrium.

Most of the games we have considered so far (the Prisoners' Dilemma, Matching Pennies, the Battle of the Sexes) have been generic. The following is an example of a non-generic game.

Example 4.38

Consider the game

$$
\begin{array}{c|c|c|c|}
 & \multicolumn{3}{c}{P_2} \\
 & L & M & R \\
\hline
U & 10, 0 & 5, 1 & 4, -2 \\
\hline
D & 10, 1 & 5, 0 & 1, -1 \\
\hline
\end{array}
$$

P_1

This game is non-generic because (D, L) is obviously a Nash equilibrium, but player 1 would get the same payoff by playing U rather than D (against L). Similarly, (U, M) is obviously a Nash equilibrium, but player 1 would get the same payoff by playing D rather than U (against M).

Theorem 4.39 (Oddness Theorem)

All generic games have an odd number of Nash equilibria.

Remark 4.40

A formal proof of the oddness theorem is rather difficult. Figure 4.1 shows the best responses for the Battle of the Sexes game. The best response for player 1 meets the best response for player 2 in three places. These are the Nash equilibria. Drawing similar diagrams for other generic games supports the truth of this theorem (at least for games between two players, each with two pure strategies).

---

[7] So this is *not* an affine transformation.

4.7 Classification of Games

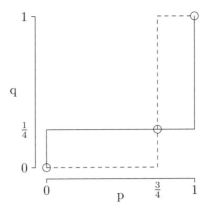

Figure 4.1 Battle of the Sexes. The best responses for player 1 are shown by a solid line and those for player 2 by a dotted line. Where they meet are the Nash equilibria (circled).

In contrast, the number of Nash equilibria in a non-generic games is (usually) infinite.

Example 4.41

Consider the game shown in Example 4.38. Define $\sigma_1 = (p, 1-p)$ and $\sigma_2 = (q, r, 1-q-r)$. Then

$$\pi_1(\sigma_1, \sigma_2) = 1 + 9q + 4r + 3p(1-q-r)$$
$$\pi_2(\sigma_1, \sigma_2) = -(1+p) + 2q + r(1+2p)$$

The best responses are

$$\hat{\sigma}_1 = \begin{cases} (1,0) & \text{if } q+r < 1 \\ (x, 1-x) \text{ with } x \in [0,1] & \text{if } q+r = 1. \end{cases}$$

$$\hat{\sigma}_2 = \begin{cases} (1,0,0) & \text{if } p < \frac{1}{2} \\ (0,1,0) & \text{if } p > \frac{1}{2} \\ (y, 1-y, 0) \text{ with } y \in [0,1] & \text{if } p = \frac{1}{2}. \end{cases}$$

So the Nash equilibria are

1. $\sigma_1^* = (x, 1-x)$ with $x \in [0, \frac{1}{2})$ and $\sigma_2^* = (1,0,0)$
2. $\sigma_1^* = (x, 1-x)$ with $x \in (\frac{1}{2}, 1]$ and $\sigma_2^* = (0,1,0)$
3. $\sigma_1^* = (\frac{1}{2}, \frac{1}{2})$ and $\sigma_2^* = (y, 1-y, 0)$ with $y \in [0,1]$

Note that the strictly dominated strategy $R \equiv (0, 0, 1)$ is not included in any of these Nash equilibria.

Exercise 4.10

Find all the Nash equilibria for the following non-generic games. Draw the best response graphs for the first game.

(a)

		P_2	
		C	D
P_1	A	6, 0	5, 3
	B	6, 1	0, 0

(b)

		P_2		
		B	F	H
P_1	G	5, 0	$-1, 1$	2, 0
	J	5, 3	$-2, 3$	2, 3

Sometimes, however, the number of Nash equilibria in a non-generic game may be finite and even.

Exercise 4.11

Consider the following game. Find all the Nash equilibria for every value of $\lambda \in (-\infty, +\infty)$.

		P_2	
		L	R
P_1	U	λ, λ	1, 1
	D	1, 1	2, 2

4.7.3 Zero-sum Games

As its name suggests, a *zero-sum game* is one in which the payoffs to the players add up to zero. For example, the game "Matching Pennies" is a zero sum game: if the first player uses a strategy $\sigma_1 = (p, 1-p)$ and the second uses $\sigma_2 = (q, 1-q)$ then their payoffs are

$$\begin{aligned}
\pi_1(\sigma_1, \sigma_2) &= pq - p(1-q) + (1-p)q - (1-p)(1-q) \\
&= (2q-1)(2p-1) \\
&= -\pi_2(\sigma_1, \sigma_2)
\end{aligned}$$

In such games the interests of the players are, therefore, exactly opposed: one only wins what the other loses. This is in contrast to many other games – such as the Prisoners' Dilemma in which both players end up wining (or losing) the same amount.

Zero-sum games were the first type of game to be studied formally. At that time, the concept of a Nash equilibrium did not exist, and games were solved

4.7 Classification of Games

by finding what was referred to as the "minimax" or ("maximin") solution. Fortunately, the minimax solution is just the Nash equilibrium for a zero-sum game. Let us define $\pi(\sigma_1, \sigma_2) = \pi_1(\sigma_1, \sigma_2)$, so $\pi_2(\sigma_1, \sigma_2) = -\pi(\sigma_1, \sigma_2)$ (in a zero-sum game). Then the Nash equilibrium conditions

$$\pi_1(\sigma_1^*, \sigma_2^*) \geq \pi_1(\sigma_1, \sigma_2^*) \quad \forall \sigma_1 \in \Sigma_1$$
$$\text{and} \quad \pi_2(\sigma_1^*, \sigma_2^*) \geq \pi_2(\sigma_1^*, \sigma_2) \quad \forall \sigma_2 \in \Sigma_2$$

can be rewritten as

$$\pi(\sigma_1^*, \sigma_2^*) = \max_{\sigma_1 \in \Sigma_1} \pi(\sigma_1, \sigma_2^*)$$
$$\text{and} \quad \pi(\sigma_1^*, \sigma_2^*) = \min_{\sigma_2 \in \Sigma_2} \pi(\sigma_1^*, \sigma_2).$$

(Remember that, to maximise their own payoff, the second player must *minimize* the first player's payoff.) By noting that each player should play a best response to the other's strategy, these two conditions can be combined

$$\pi(\sigma_1^*, \sigma_2^*) = \max_{\sigma_1 \in \Sigma_1} \pi(\sigma_1, \sigma_2^*)$$
$$= \max_{\sigma_1 \in \Sigma_1} \min_{\sigma_2 \in \Sigma_2} \pi(\sigma_1, \sigma_2)$$

or, equivalently

$$\pi(\sigma_1^*, \sigma_2^*) = \min_{\sigma_2 \in \Sigma_2} \pi(\sigma_1^*, \sigma_2)$$
$$= \min_{\sigma_2 \in \Sigma_2} \max_{\sigma_1 \in \Sigma_1} \pi(\sigma_1, \sigma_2).$$

Exercise 4.12

"Ace-King-Queen" is a simple card game for two players, which is played as follows. The players each bet a stake of $5. Each player then chooses a card from the set {Ace, King, Queen} and places it face down on the table. The cards are turned over simultaneously, and the winner of the hand is decided by the following rules: an "Ace" (A) beats a "King" (K); a "King" beats a "Queen" (Q); and a "Queen" beats an "Ace". The winning player takes the $10 in the pot. If both players choose the same card (both A, both K, or both Q), the game is drawn and the $5 stake is returned to each player. What is the unique Nash equilibrium for this game?

Theorem 4.42

A generic zero-sum game has a unique solution.

Proof

See Von Neuman & Morgenstern (1953). □

Exercise 4.13

Consider the game shown below. Show that if the game is generic (i.e., $a \neq b$, $a \neq c$, etc.), then there is a unique Nash equilibrium. [Hint: there are sixteen possible cases to consider.]

	C	D
A	$a, -a$	$b, -b$
B	$c, -c$	$d, -d$

4.8 Games with n-players

The extension of the theory to games with more than two players is straightforward, if notationally baroque. Let us label the players by $i \in \{1, 2, \ldots, n\}$ Each player has a set of pure strategies $\mathbf{S}_i$ and a corresponding set of mixed strategies $\mathbf{\Sigma}_i$. The payoff to player i depends on a list of strategies $\sigma_1, \sigma_2, \ldots, \sigma_n$ – one for each player. For the definition of a Nash equilibrium, we will need to separate out the strategy for each of the players, so we denote by σ_{-i} the list of strategies used by all the players except the i-th player.

Example 4.43

Consider a game with three players. The payoffs to each player can be written as:

$$\pi_1(\sigma_1, \sigma_{-1}) \equiv \pi_1(\sigma_1, \sigma_2, \sigma_3)$$
$$\pi_2(\sigma_2, \sigma_{-2}) \equiv \pi_2(\sigma_1, \sigma_2, \sigma_3)$$
$$\pi_3(\sigma_3, \sigma_{-3}) \equiv \pi_3(\sigma_1, \sigma_2, \sigma_3).$$

Suppose player i uses a mixed strategy σ_i which specifies playing pure strategy $s \in \mathbf{S}_i$ with probability $p_i(s)$. Payoffs for mixed strategies are then calculated from the payoff table with entries $\pi_i(s_1, \ldots, s_n)$ by

$$\pi_i(\sigma_i, \sigma_{-i}) = \sum_{s_1 \in \mathbf{S}_1} \cdots \sum_{s_n \in \mathbf{S}_n} p_1(s_1) \cdots p_n(s_n) \pi_i(s_1, \ldots, s_n)$$

4.8 Games with n-players

	P_2	
A	L	R
P_1 U	$1,1,0$	$2,2,3$
P_1 D	$2,2,3$	$3,3,0$

	P_2	
B	L	R
P_1 U	$-1,-1,2$	$2,0,2$
P_1 D	$0,2,2$	$1,1,2$

Figure 4.2 A representation of the three player game from example 4.45.

Definition 4.44

A Nash equilibrium in a n-player game is a list of mixed strategies $\sigma_1^*, \sigma_2^*, \ldots, \sigma_n^*$ such that
$$\sigma_i^* \in \underset{\sigma_i \in \Sigma_i}{\operatorname{argmax}} \pi_i(\sigma_i, \sigma_{-i}^*) \qquad \forall i \in \{1, 2, \ldots, n\}$$

Example 4.45

Consider a static three-player game where the first player chooses between U and D, the second player chooses between L and R, and the third player chooses between A and B. Instead of trying to draw a three-dimensional payoff table, we represent this game by a pair of payoff tables such as the ones shown in Figure 4.2. (We can interpret this as player 3 choosing the game that players 1 and 2 have to play, so long as we remember that players 1 and 2 do not know which of the payoff tables player 3 has chosen.) We can find a Nash equilibrium for the game with the payoffs shown in Figure 4.2 as follows. First, suppose that player 3 chooses A. Then the best responses for players 1 and 2 are the strategies $\hat{\sigma}_1 = \hat{\sigma}_2 = (0, 1)$. However, we do not have a Nash equilibrium because choosing A is not player 3's best response to this pair of strategies. Now suppose that player 3 chooses B. Then the best responses for players 1 and 2 are the strategies $\hat{\sigma}_1 = \hat{\sigma}_2 = (\frac{1}{2}, \frac{1}{2})$. Because player 3 would get a payoff of $\frac{3}{2}$ if he switches to A, we have a Nash equilibrium $(\sigma_1^*, \sigma_2^*, \sigma_3^*)$ with

$$\sigma_1^* = (\frac{1}{2}, \frac{1}{2}) \qquad \sigma_2^* = (\frac{1}{2}, \frac{1}{2}) \qquad \sigma_3^* = (0, 1)$$

Exercise 4.14

Represent the game from Example 4.45 by a pair of payoff tables "chosen" by player 2. Confirm that the game has the same Nash equilibrium when represented in this way. [Hint: show that there are no pure strategy Nash equilibria, then use the Equality of Payoffs theorem to find a Nash equilibrium involving mixed strategies.]

5
Finite Dynamic Games

5.1 Game Trees

So far we have considered static games in which decisions are assumed to be made simultaneously (or, at least, in ignorance of the choices made by the other players). However, there are many situations of interest in which decisions are made at various times with at least some of the earlier choices being public knowledge when the later decisions are being made. These games are called *dynamic games* because there is an explicit time-schedule that describes when players make their decisions.

Dynamic games can be represented by a game tree – the so-called *extensive form* – which is an extension of the decision tree used in (single-person) decision theory. The times at which decisions are made are shown as small, filled circles. Leading away from these *decision nodes* is a branch for every action that could be taken at that node. When every decision has been made, one reaches the end of one path through the tree. At that point, the payoffs for following that path is written. We will use the convention that the first payoff in each pair is for the player who moves first. Time increases as one goes down the page, so the tree is drawn "upside-down".

Example 5.1 (Dinner Party Game)

Two people ("husband" and "wife") are buying items for a dinner party. The husband buys either fish (F) or meat (M) for the main course; the wife buys

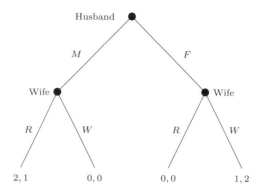

Figure 5.1 The game tree for the dinner party game of Example 5.1.

either red wine (R) or white wine (W). Both people are rather conventional and prefer red wine with meat and white wine with fish, rather than either of the opposite combinations, which are equally displeasing. However, the husband prefers meat over fish, while the wife prefers fish over meat. We can represent these preferences as utility-based payoffs:

$$\pi_h(M,R) = 2 \quad \pi_h(F,W) = 1 \quad \pi_h(F,R) = \pi_h(M,W) = 0$$

$$\pi_w(M,R) = 1 \quad \pi_w(F,W) = 2 \quad \pi_w(F,R) = \pi_w(M,W) = 0$$

where the payoffs for the husband have been given a subscript h and those for the wife a subscript w. So far the description of the game has been no different from that of a static game. Let us now assume that the husband buys the main course and tells his wife what was bought; his wife then buys some wine. The game tree for this game is shown in Figure 5.1.

What is the solution of the dinner party game? The obvious way to solve this game is by backward induction (i.e., to work backwards through the game tree). Recall that the husband tells his wife whether fish or meat has been purchased. So when she makes her decision about the wine, she knows what main dish her husband will be cooking. If the husband has bought fish, then his wife will buy white wine (because this gets her a payoff = 2, rather than a payoff = 0 for having red wine with fish). On the other hand, if the husband has bought meat, then his wife will buy red wine (a payoff = 1 rather than a payoff = 0 for white wine with meat). So if the husband buys fish, then his wife will buy white wine and he will get a payoff = 1. On the other hand, if the husband buys meat, then his wife will buy red wine and he will get a payoff

5.2 Nash Equilibria

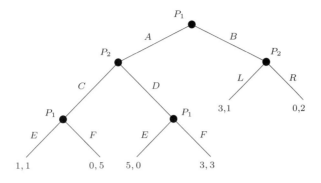

Figure 5.2 Game tree for Exercise 5.1.

$= 2$. So the husband prefers to buy meat and the dinner party will consist of meat and red wine (and, hopefully, some other items).

Exercise 5.1

Find a solution using backward induction for the game shown in Figure 5.2.

5.2 Nash Equilibria

Is the solution we have just derived a Nash equilibrium? To answer this question, we have to determine what the strategies are for each player. The action sets for each player are $\mathbf{A}_h = \{M, F\}$ and $\mathbf{A}_w = \{R, W\}$. The set of pure strategies available to the husband is the same as his action set: $\mathbf{S}_h = \{M, F\}$. However, the wife has *four* possible pure strategies: $\mathbf{S}_w = \{RR, RW, WR, WW\}$ where[1]

$$
\begin{aligned}
RR &\equiv \text{"R if her husband chooses M and R if he chooses F"} \\
RW &\equiv \text{"R if her husband chooses M and W if he chooses F"} \\
WR &\equiv \text{"W if her husband chooses M and R if he chooses F"} \\
WW &\equiv \text{"W if her husband chooses M and W if he chooses F"}
\end{aligned}
$$

So in strategic (or normal) form the game has the following payoff table (with best responses underlined).

---

[1] The wife's strategy set clearly illustrates the difference between actions and strategies – a distinction that cannot be made in static games.

		Wife			
		RR	RW	WR	WW
Husband	M	<u>2</u>,<u>1</u>	<u>2</u>,<u>1</u>	0,0	0,0
	F	0,0	1,<u>2</u>	0,0	<u>1</u>,<u>2</u>

Clearly there are three pure strategy Nash equilibria: (M, RR), (M, RW), and (F, WW). Any pair (M, σ_2^*) where σ_2^* assigns probability p to RR and $1-p$ to RW is also a Nash equilibrium. So the solution we found by backward induction is a Nash equilibrium but there are many others.

Although there are many Nash equilibria, not all are equally believable if we consider what they imply for the behaviour of one of the players. Consider the wife's strategy WW. This corresponds to the wife telling her husband that if he buys meat she will, nevertheless, buy white wine. In response to this announcement her husband should buy the fish that his wife prefers because, if he does not, he will get one of his least preferred outcomes (white wine with meat). However, if her husband *has* bought meat, then the wife should buy red wine when she comes to her decision. This is because she prefers red wine compared to white wine when the dinner is based on meat. So the husband should not believe his wife when she threatens to buy white wine if he buys meat. In other words, the Nash equilibrium (F, WW) relies on the wife threatening to choose an option she would not take if she were faced with the decision of choosing a wine to go with meat.

Consider, now, the wife's strategies RR and σ_2^*. Both of these specify that the wife will buy red wine if her husband buys meat, which is alright. However, the first says that the wife would *definitely* buy red wine if her husband chose fish and the second that she *may* (if $p > 0$) buy red wine to go with fish. Neither of these is believable because the wife *definitely* prefers white wine with fish. (Note that neither of these strategies can be called a "threat" because the husband gets his most preferred outcome of meat and red wine in any case.)

The Nash equilibria found from the strategic form don't all seem to capture the essence of the dynamic game, because the order of the decisions is suppressed. Rather, a subset of the Nash equilibria – the ones found by backward induction on a game tree – seem more reasonable than the others when the time structure is taken into account.

Exercise 5.2

Consider the game tree shown in Figure 5.3. Solve this game by backward induction. Give the strategic form of the game and find all the pure strategy equilibria.

5.3 Information Sets

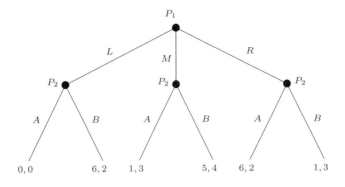

Figure 5.3 Game tree for Exercise 5.2.

5.3 Information Sets

The difference between static games and dynamic games is not that the former can be represented in strategic form and the latter by a game tree. After all, we have just taken a dynamic game and represented it in strategic form in order to find all the Nash equilibria and, as we shall see, static games have a game tree representation too.

The real distinction between static and dynamic games is what is known by the players when they make their decisions. In the dinner party game, the wife knew whether her husband had bought meat or fish when the time came for her to choose between red wine and white wine. We formally specify what is known to a player by giving their information set.

Definition 5.2

An *information set* for a player is a set of decision nodes in a game tree such that:

1. the player concerned (and no other) is making a decision;
2. the player does not know which node has been reached (only that it is one of the nodes in the set).

Note that the second part of this definition requires that a player must have the same choices at all nodes included in an information set.

Example 5.3

Consider a static version of the dinner party game. Suppose that, although the husband chooses first, his wife does not know what main course ingredient has been bought when she is trying to choose a wine. A game tree for this game is shown in Figure 5.4(a) where the dotted line joining the wife's two decision nodes represents the fact that she does not know which node she is at (i.e., both her decision nodes constitute an information set). The strategic form of this game has the payoff table shown below.

		Wife	
		R	W
Husband	M	2, 1	0, 0
	F	0, 0	1, 2

Note that the wife only has two pure strategies in this case because she cannot condition her actions on her husband's behaviour (because she doesn't know it). Clearly, because the husband does not know what his wife will choose, she could choose first (and not tell him what wine she has bought) without changing the game. Therefore, a second possible game tree for this game is the one shown in Figure 5.4(b).

Exercise 5.3

Draw two different trees for the static game below. Can any solutions be found by backward induction?

		P_2		
		L	M	R
P_1	U	4, 3	2, 7	0, 4
	D	5, 5	5, −1	−4, −2

Exercise 5.4

Draw the game tree for a version of the Prisoners' Dilemma where one prisoner knows what the other has done. Is the outcome affected by the decisions being sequential rather than simultaneous?

5.4 Behavioural Strategies

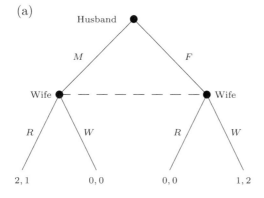

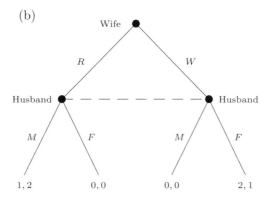

Figure 5.4 Two possible game trees for the static version of the dinner party game of Exercise 5.3. The dotted lines between decision nodes indicate that those nodes belong to the same information set.

5.4 Behavioural Strategies

In general, a dynamic game a player may encounter an information set containing two or more decision nodes. At this point, the player does not have complete information about the behaviour of their opponent – the two players are making "simultaneous decisions". The presence of such information sets means that we must allow for the possibility that players will randomise as they might do in a single-decision, static game.

Example 5.4

Consider a game in which player 1 chooses between actions A and B. If A is chosen, then players 1 and 2 play a game of "matching pennies". If player 1 chooses B, then player 2 chooses L or R. The tree for this game is shown in Figure 5.5(a). Let us try to solve this game by an extended version of backward induction. Let σ be the strategy "Play H with probability $\frac{1}{2}$", then (σ, σ) is the unique Nash equilibrium for matching pennies. If we assume that the players will indeed play the matching pennies game in this way, we can replace this part of the tree with the expected payoffs for the two players – in this case $(0,0)$. We then have the truncated game tree shown in Figure 5.5(b). On the right-hand side of the tree player 2 should obviously choose R, which leads to the truncated game tree shown in Figure 5.5(c). So, at the start of the game, if player 1 chooses A they get an expected payoff of zero. On the other hand, if they choose B, they get a payoff of -1. So the backward induction solution is that player 1 should use the strategy "A then σ" and player 2 should use "σ if A, R if B". We can shorten this solution without ambiguity to $(A\sigma, \sigma R)$.

As we saw in Section 2.3, there are two ways of defining a randomising strategy. The randomising strategy we found in the previous example is known as a *behavioural strategy*. In a behavioural strategy, the opportunity for randomisation (by the appropriate player) occurs at each information set. In working backwards through the game tree we found a best response at each information set so the end result is an equilibrium in behavioural strategies. The alternative is known as a *mixed strategy*, which is formed by taking weighted combinations of pure strategies (see Section 4.2)

$$\sigma = \sum_{s \in \mathbf{S}} p(s)s \quad \text{with} \quad \sum_{s \in \mathbf{S}} p(s) = 1 \ .$$

It is randomising strategies defined in the second way that appear in the definition of a Nash equilibrium. When we wish to distinguish between the two sorts of strategy, we will denote a behavioural strategy by the symbol β.

In Section 5.2, we saw that the equilibrium in behavioural strategies was equivalent to a Nash equilibrium of the strategic form game in a specific example. The next theorem shows that for any equilibrium in behavioural strategies there is a Nash equilibrium in mixed strategies that gives the same payoffs to both players.

Theorem 5.5

Let (β_1^*, β_2^*) be an equilibrium in behavioural strategies. Then there exist mixed strategies σ_1^* and σ_2^* such that

5.4 Behavioural Strategies

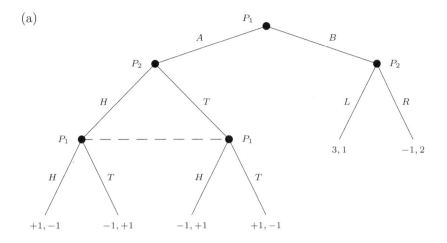

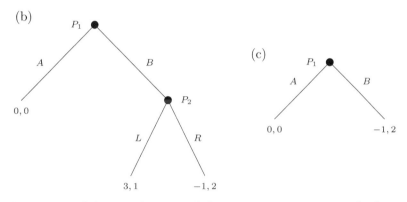

Figure 5.5 Solution of a general dynamic game containing both sequential and simultaneous decisions by backward induction (a) Original game. (b) First truncation. (c) Second truncation. (See Example 5.4 for a description of the procedure.)

(a) $\pi_i(\sigma_1^*, \sigma_2^*) = \pi_i(\beta_1^*, \beta_2^*)$ for $i = 1, 2$ and

(b) the pair of strategies (σ_1^*, σ_2^*) is a Nash equilibrium.

Proof

(a) Consider one of the players. For any fixed strategy of the other player, it follows from Theorem 2.16 (by replacing "decision nodes" with "information sets") that every behavioural strategy has a mixed strategy representation.[2] Because the two representations assign the same weight to each path through the game tree, it follows that $\pi_i(\sigma_1^*, \sigma_2^*) = \pi_i(\beta_1^*, \beta_2^*)$ for $i = 1, 2$.

(b) Suppose that, although (β_1^*, β_2^*) is an equilibrium in behavioural strategies, (σ_1^*, σ_2^*) is not a Nash equilibrium. Then one of the players must have an alternative strategy that yields a higher payoff. Without loss of generality, we will assume this is player 1 and call this strategy σ_1'. Because this strategy has a different payoff against σ_2^*, its behavioural representation must be different from β_1^*. Let us call it β_1'. Then

$$\begin{aligned}
\pi_1(\beta_1', \beta_2^*) &= \pi_1(\sigma_1', \sigma_2^*) \\
&> \pi_1(\sigma_1^*, \sigma_2^*) \\
&= \pi_1(\beta_1^*, \beta_2^*)
\end{aligned}$$

which contradicts the assumption that (β_1^*, β_2^*) is an equilibrium in behavioural strategies. □

Now that we have shown that equilibria in behavioural strategies are equivalent to Nash equilibria, we can drop the distinction between behavioural and mixed strategies and denote an arbitrary strategy by σ.

Exercise 5.5

Find the strategic form of the game from Example 5.4. Find mixed strategies σ_1^* and σ_2^* that give both players the same payoff they achieve by using the behavioural strategies found by backward induction. Show that the pair (σ_1^*, σ_2^*) is a Nash equilibrium.

Exercise 5.6

A firm (the "Incumbent") has a monopoly in a market worth £6 million. A second firm (the "Newcomer") is thinking of entering this market. If the Newcomer does enter the market, the Incumbent can either do nothing or start a price war. The cost of a price war is £2 million to each firm. If the Newcomer enters then the two firms share the market equally. If the Newcomer does not enter then its next best option provides an income of £2 million. (a) Draw a game tree for this situation and find an equilibrium in behavioural strategies. (b) Construct the Strategic Form of this game and find all the Nash equilibria.

---

[2] We assume that the players have perfect recall – that is, they do not forget the decisions they have made in the past. This ensures that each player makes a unique sequence of decisions to arrive at any particular information set.

5.5 Subgame Perfection

Using the extended form of backward induction to eliminate "unreasonable" Nash equilibria finds what are known as *subgame perfect* Nash equilibria. In this section, we give a formal definition of subgame perfection.

Definition 5.6

A *subgame* is a part (sub-tree) of a game tree that satisfies the following conditions.

1. It begins at a decision node (for any player).
2. The information set containing the initial decision node contains no other decision nodes. That is, the player knows all the decisions that have been made up until that time.
3. The sub-tree contains all the decision nodes that follow the initial node (and no others).

Example 5.7

In the sequential decision dinner party game of Figure 5.1, the subgames are (i) the parts of the game tree beginning at each of the wife's decision nodes and (ii) the whole game tree.

Example 5.8

The only subgame of the "simultaneous" decision dinner party game (in either version of the game tree shown in Figure 5.4) is the whole game.

Definition 5.9

A *subgame perfect Nash equilibrium* is a Nash equilibrium in which the behaviour specified in every subgame is a Nash equilibrium for the subgame. Note that this applies even to subgames that are *not* reached during a play of the game using the Nash equilibrium strategies.

Example 5.10

In the dinner party game of Example 5.1, the Nash equilibrium (M, RW) is a subgame perfect Nash equilibrium because (i) the wife's decision in response

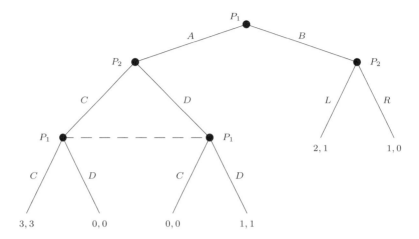

Figure 5.6 A dynamic game with multiple subgame perfect Nash equilibria. See Example 5.11 for a description of the solution.

to a choice of meat is to choose red wine, which is a Nash equilibrium in that subgame; (ii) the wife's decision in response to a choice of fish is to choose white wine (a Nash equilibrium in that subgame); and (ii) the husband's decision is to choose meat, which (together with his wife's strategy of RW, constitutes a Nash equilibrium in the entire game. However, the Nash equilibrium (F, WW) is not subgame perfect because it specifies a behaviour (choosing W) that is not a Nash equilibrium for the subgame beginning at the wife's decision node following a choice of meat by her husband.

It follows from the definition of a subgame perfect Nash equilibrium that any Nash equilibrium that is found by backward induction is subgame perfect. If a simultaneous decision subgame occurs, then all possible Nash equilibria of this subgame may appear in some subgame perfect Nash equilibrium for the whole game.

Example 5.11

Consider the game described by the game tree in Figure 5.6. The simultaneous decision subgame has three Nash equilibria: (C, C), (D, D), and a mixed strategy equilibrium (σ_1^*, σ_2^*) giving each player a payoff of $\frac{1}{4}$. So the subgame perfect Nash equilibria are (AC, CL), (BD, DL), and $(B\sigma_1^*, \sigma_2^* L)$.

5.6 Nash Equilibrium Refinements

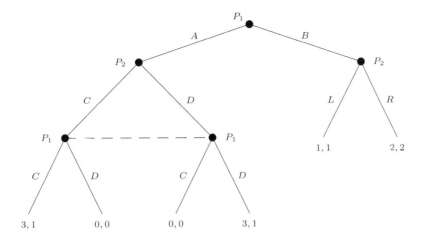

Figure 5.7 Game tree for Exercise 5.7.

Theorem 5.12

Every finite dynamic game has a subgame perfect Nash equilibrium.

Proof

The result follows immediately from Definition 5.9 together with Nash's theorem. □

Exercise 5.7

Find all the subgame perfect Nash equilibria for the game shown in Figure 5.7.

Exercise 5.8

Find all the subgame perfect Nash equilibria of the game shown in Figure 5.8.

5.6 Nash Equilibrium Refinements

Subgame perfection is one of many proposed *Nash equilibrium refinements*. These attempt to supplement the definition of a Nash equilibrium with extra

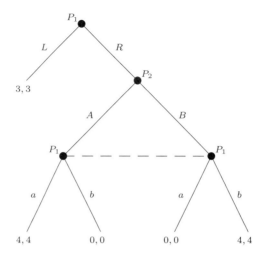

Figure 5.8 Game tree for Exercise 5.8.

conditions in order to reduce the number of equilibria to a more "reasonable" set. Ideally, the number of equilibria would be reduced to one, and that equilibrium would then be considered *the* solution of the game. There are two problems with this approach. First, the definition of "reasonable" varies according to the situation being modelled. Second, the number of equilibria that satisfy the refinement conditions is rarely just one: often several equilibria remain – see Example 5.11. Moreover, while subgame perfect equilibria always exist, other types of refinement may lead to some games having no equilibria that satisfy the additional conditions.

Subgame perfection tries to select particular equilibria as being more reasonable by moving backwards through the game tree. An alternative approach, called *forward induction*, moves forward through the tree. Let us look again at Example 5.11. The "problem" with subgame perfection in that game is that it does not provide a way to select between the three possible Nash equilibrium behaviours in the simultaneous decision subgame. However, if play has reached that subgame, player 2 could reasonably assume that player 1 will use C because it is only the (C, C) equilibrium that will result in a payoff greater than 2 (which player 1 could have received by using B at the beginning). Player 1, realising that their opponent will reach this conclusion, is then confident of receiving a payoff of 3 for choosing A at the beginning and, therefore, chooses that action instead of B. Thus the equilibrium supported by forward induction is (AC, CL).

5.6 Nash Equilibrium Refinements

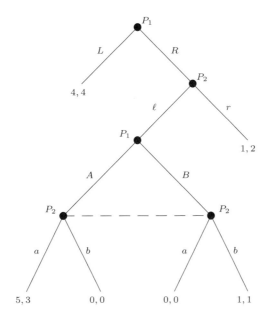

Figure 5.9 Game tree for Exercise 5.9.

Exercise 5.9

Consider the game shown in Figure 5.9. Find all the subgame perfect Nash equilibria. Which of these equilibria is supported by a forward induction argument?

A problem that is common to both subgame perfection and forward induction is that they assume the players will behave rationally (i.e., select Nash equilibrium behaviours that satisfy all the supplementary conditions that have been deemed reasonable) in parts of the game tree that would not be reached if the players act as prescribed by the equilibrium. But if those parts of the game tree will only be reached as a consequence of *irrational* behaviour by one or more players, why should we – or, indeed, the players themselves – assume that rational behaviour will reassert itself at that point?

Example 5.13

Consider the game shown in Figure 5.10. The unique subgame perfect Nash equilibrium is one in which both players will play L at every opportunity. Therefore, we (and the players) should expect player 1 to use L at the beginning of the game – at which point the game ends. Suppose that, contrary to this

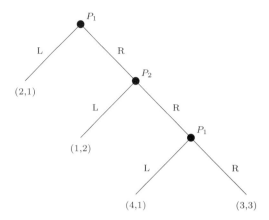

Figure 5.10 A game that illustrates a problem with subgame perfection. The unique subgame perfect Nash equilibrium leads to player 1 using L at the beginning of the game. But what should player 2 do if player 1 behaves irrationally and gives them the opportunity to make a decision?

expectation, player 1 uses R and consequently player 2 gets to make a decision. What should player 2 do? The backward induction argument, which is based on the assumption that player 1 will behave rationally in the future, would suggest that player 2 should use L at this point. But player 1 has already behaved irrationally once, so perhaps they will do so again. This argument suggests that player 2 may be better off choosing R, thus giving player 1 the opportunity to use R again.

The problem is that we are attempting to analyse irrational behaviour on the basis that the players are rational. Although this appears to be a problem that is impossible to solve, there is one way to cut the Gordian knot. This is to assume that the players are perfectly rational in their intentions but that they make mistakes in the execution of those intentions. In other words, the unexpected behaviour is only apparently irrational. Think of a chess player about to move a piece. Two legal moves are available, one of which is better than the other. The player picks up the piece and moves it towards the better of the finishing positions. However, at the last minute, their hand trembles and they place the piece in the "wrong" square. Using this analogy, Nash equilibria that remain when the possibility of (small) mistakes is taken into account are called *trembling hand Nash equilibria*. Typically the probability of a mistake is characterised by a number ε and the limit as $\varepsilon \to 0$ is taken.

This characterisation of unexpected behaviour implies that a mistake by a

5.6 Nash Equilibrium Refinements

player at one time does not make it more likely that the same player will make another mistake in the future – the "trembles" are uncorrelated. This supports the subgame perfect Nash equilibrium in the game shown in Figure 5.10: player 2 should use L if they get the opportunity, so long as the probability of making a mistake is small enough.

Exercise 5.10

Consider the game from Example 5.13. Let the probability that player 1 makes a mistake be ε. Find an $\bar{\varepsilon}$ such that for all $\varepsilon < \bar{\varepsilon}$ player 2 should use L if given the opportunity.

6
Games with Continuous Strategy Sets

6.1 Infinite Strategy Sets

For ease of exposition, most of this book is devoted to models in which players have discrete and finite strategy sets. However, several classic games describe situations in which the players do not choose actions from a discrete set; instead their pure strategy sets are subsets of the real line. In this chapter, we give a few examples to show how the concepts of game theory are easily extended to such cases. Economic models of a duopoly provide examples with pure-strategy Nash equilibria, and the so-called War of Attrition has an equilibrium involving mixed strategies.

Suppose the pure strategy (action) sets are a subset of the real line $[a, b]$. A pure strategy is then a choice $x \in [a, b]$ and a mixed strategy is defined by giving a function $p(x)$ such that the probability that the choice lies between x and $x+dx$ is $p(x)dx$. The existence of Nash equilibria for games with continuous pure-strategy sets was proved independently by Debreu, Glicksburg, and Fan in 1952 (see Myerson (1991) or Fudenberg & Tirole (1993) for details).

6.2 The Cournot Duopoly Model

A duopoly is a market in which two firms compete to supply the same set of customers with the same product. There are three classic duopoly models:

the Cournot duopoly, the Bertrand duopoly, and the Stackelberg duopoly – all named after their originators. In the Bertrand duopoly, the amount of the product consumed is determined by the price at which it is sold and the two firms have to decide simultaneously the price at which they will try to sell their stock. The firm that sets the lower price "captures the market" and the firm that sets the higher price sells nothing. A simple analysis of this situation using the elimination of dominated strategies shows that the firms should set a price that exactly matches their costs of production (otherwise the other firm could undercut their price). In the Cournot and Stackelberg duopoly models, the two firms have to decide how much of their product to manufacture, and the price at which the product is sold is determined by the total amount made. In the Cournot model, the decisions are made simultaneously while in the Stackelberg model the decisions are made sequentially with the decision of the first firm being public knowledge. Economists sometimes call the solutions of these three models a "Bertrand equilibrium", a "Cournot equilibrium", and a "Stackelberg equilibrium". However, they are all just Nash equilibria of their respective models, with the equilibrium in the Stackelberg case being subgame perfect.

Consider two firms competing for a market by making some infinitely divisible product, such as petroleum. Cournot's model is based on allowing the firms to choose how much of the product they make, so the set of actions for each firm is a range of quantities q_i which it could produce. Because the product is infinitely divisible, this action set is continuous.

If Firm i produces an amount q_i of the product, the total amount produced is $Q = q_1 + q_2$. The market price of the product is assumed to depend on the total supply:

$$P(Q) = \begin{cases} P_0 \left(1 - \frac{Q}{Q_0}\right) & \text{if } Q < Q_0 \\ 0 & \text{if } Q \geq Q_0. \end{cases}$$

So the market price drops from a maximum of P_0 when the product is very scarce to zero when a quantity Q_0 is is produced. The production costs are assumed to be $C(q_i) = cq_i$ (i.e., there are no fixed costs and the cost of making a unit of the product is the same for each firm). The payoff for each firm is given by the profit that it makes in a market determined by the behaviour of both firms. The payoff to Firm i is, therefore,

$$\pi_i(q_1, q_2) = q_i P(Q) - cq_i.$$

Notice that it certainly makes no sense for either firm to produce a quantity greater than Q_0, because that would certainly lead to a loss rather than a profit. Consequently, we can restrict the action set to the range $[0, Q_0]$.

We begin by finding the best response for Firm 1 against every possible production quantity that Firm 2 could choose. The best response is to choose

6.2 The Cournot Duopoly Model

a production quantity $\hat{q}_1$ that maximises the profit for Firm 1, given a value of q_2. So we solve

$$\frac{\partial \pi_1}{\partial q_1}(\hat{q}_1, q_2) = 0$$

to find

$$\hat{q}_1 = \frac{Q_0}{2}\left(1 - \frac{q_2}{Q_0} - \frac{c}{P_0}\right).$$

To check that this is really a best response (and not a "worst response") we calculate

$$\frac{\partial^2 \pi_1}{\partial q_1^2}(\hat{q}_1, q_2) = -\left(\frac{P_0}{Q_0}\right)$$
$$< 0.$$

We also need to confirm that $\hat{q}_1 + q_2 \leq Q_0$, so that the firms are making a non-negative profit:

$$\begin{aligned}
\hat{q}_1 + q_2 &= \frac{Q_0}{2}\left(1 - \frac{q_2}{Q_0} - \frac{c}{P_0}\right) + q_2 \\
&= \frac{Q_0}{2} + \frac{q_2}{2} - \frac{cQ_0}{2P_0} \\
&\leq \frac{Q_0}{2} + \frac{Q_0}{2} - \frac{cQ_0}{2P_0} \\
&= Q_0\left(1 - \frac{c}{2P_0}\right) \\
&< Q_0.
\end{aligned}$$

Similarly, we find the best response to a choice of q_1 is for Firm 2 to produce

$$\hat{q}_2 = \frac{Q_0}{2}\left(1 - \frac{q_1}{Q_0} - \frac{c}{P_0}\right).$$

A pure strategy Nash equilibrium is a pair (q_1^*, q_2^*), each of which is a best response to the other. Such a pair can be found by solving the simultaneous equations

$$\begin{aligned}
q_1^* &= \frac{Q_0}{2}\left(1 - \frac{q_2^*}{Q_0} - \frac{c}{P_0}\right) \\
q_2^* &= \frac{Q_0}{2}\left(1 - \frac{q_1^*}{Q_0} - \frac{c}{P_0}\right).
\end{aligned}$$

The solution is

$$q_1^* = q_2^* = \frac{Q_0}{3}\left(1 - \frac{c}{P_0}\right) \equiv q_c^*$$

where we have defined q_c^* as the value of the quantity chosen at the equilibrium by each firm. At this equilibrium the payoff to each firm is

$$\begin{aligned}\pi_1(q_c^*, q_c^*) &= \pi_2(q_c^*, q_c^*) \\ &= q_c^* P(2q_c^*) - cq_c^* \\ &= \frac{Q_0 P_0}{9}\left(1 - \frac{c}{P_0}\right)^2\end{aligned}$$

Let us compare this competitive equilibrium with the situation that holds under a monopoly. A monopolist maximises

$$\pi_m(q) = qP(q) - cq$$

and the optimal strategy for a monopolist is, therefore,

$$q_m^* = \frac{Q_0}{2}\left(1 - \frac{c}{P_0}\right).$$

Because $q_m^* < 2q_c^*$, the price at which goods are sold is higher for the monopoly than it is for the two competing firms. So the model indicates that competition operates to benefit the consumer.

Suppose, instead, the two firms in the duopoly could form a cartel and agree to use the strategies

$$q_1 = q_2 = \frac{1}{2}q_m^*.$$

That is they each produce half of the optimum quantity for a monopolist. Then they would receive profits of

$$\begin{aligned}\pi_i(\tfrac{1}{2}q_m^*, \tfrac{1}{2}q_m^*) &= \tfrac{1}{2}q_m^* P(q_m^*) - \tfrac{1}{2}cq_m^* \\ &= \frac{Q_0 P_0}{8}\left(1 - \frac{c}{P_0}\right)^2\end{aligned}$$

which are greater than the Cournot payoff, and the price paid by consumers would be the same as they would pay under a monopoly. However, such collusion is unstable, because the best response to a firm producing the cartel quantity is to produce

$$\begin{aligned}\hat{q} &= \frac{Q_0}{2}\left(1 - \frac{q_m^*}{2Q_0} - \frac{c}{P_0}\right) \\ &= \frac{3}{4}q_m^* \\ &> \frac{1}{2}q_m^*.\end{aligned}$$

Note that we have not proved that cartels are impossible, only that they will not occur in situations described by the Cournot model. We will return to this

Exercise 6.1

Consider the asymmetric Cournot duopoly game where the marginal cost for Firm 1 is c_1 and the marginal cost for Firm 2 is c_2. If $0 < c_i < \frac{1}{2}P_0$ $\forall i$, what is the Nash equilibrium? If $c_1 < c_2 < P_0$ but $2c_2 > P_0 + c_1$, what is the Nash equilibrium?

Exercise 6.2

Consider the n-player Cournot game. n identical firms (i.e., identical costs) produce quantities $q_1, q_2, \ldots, q_n$. The market price is given by $P(Q) = P_0(1 - Q/Q_0)$ where $Q = \sum_{i=1}^{n} q_i$. Find the symmetric Nash equilibrium (i.e., $q_i^* = q^*$ $\forall i$). What happens to each firm's profit as $n \to \infty$?

Exercise 6.3

Two adjacent countries (labelled by $i = \{1, 2\}$) each have industries that emit pollution at a level e_i tonnes per annum. Pollution from one country has a reduced effect on the other, so that the total level of pollution in country 1 is $E_1 = e_1 + ke_2$ (where $0 < k < 1$) and the total level of pollution in country 2 is $E_2 = e_2 + ke_1$. Initially, each country produces an amount of pollution e_0. However, the parliament in each country can vote to reduce the amount of pollution that it produces at a cost of c pounds per tonne per annum. The cost to the government-funded health service in each country increases with the total level of pollution as $B_0 E_i^2$. Construct the payoffs $B_i(e_1, e_2)$ for each of the countries and determine the equilibrium level of pollution produced in each country, assuming that the parliaments vote simultaneously.

6.3 The Stackelberg Duopoly Model

In the Stackelberg model, two firms ($i = 1, 2$) are competing to sell a divisible product and must decide how much of it to produce, q_i. As in the Cournot model, we assume that the market price for the product is given by

$$P(Q) = P_0 \left(1 - \frac{Q}{Q_0}\right)$$

where $Q = q_1 + q_2$ and that the cost of a unit of production for each firm is c. Unlike the Cournot duopoly model, decisions are made sequentially: Firm

1 (termed the "market leader") decides on a quantity to produce and this decision is observed by Firm 2 (the "market follower"), which then decides on the quantity that it will produce. As usual, we assume that each firm wishes to maximise its profit and that $P_0 > c$.

We solve this game by backward induction to find a subgame perfect Nash equilibrium. We begin by finding the best response of Firm 2, $\hat{q}_2(q_1)$, for every possible choice of production quantity by Firm 1. Given that Firm 1 knows Firm 2's best response to every choice of q_1, we can find a Nash equilibrium for this game by determining the maximum payoff that Firm 1 can achieve given that Firm 2 will always use its best response to any particular choice of quantity by Firm 1.

Firm 2's profit is $\pi_2(q_1, q_2) = q_2[P(Q) - c]$ and the best response to a choice of q_1 is found by solving

$$\frac{\partial \pi_2}{\partial q_2}(q_1, \hat{q}_2) = 0$$

which gives

$$\hat{q}_2(q_1) = \frac{Q_0}{2}\left(1 - \frac{q_1}{Q_0} - \frac{c}{P_0}\right)$$

If Firm 1 chooses q_1 and Firm 2 chooses the best response $\hat{q}_2(q_1)$, Firm 1's profit is

$$\begin{aligned} \pi_1(q_1, \hat{q}_2(q_1)) &= q_1\left[P_0\left(1 - \frac{q_1 + q_2(q_1)}{Q_0}\right) - c\right] \\ &= q_1 \frac{P_0}{2}\left(1 - \frac{q_1}{Q_0} - \frac{c}{P_0}\right). \end{aligned}$$

So Firm 1 maximises its profit at

$$\hat{q}_1 = \frac{Q_0}{2}\left(1 - \frac{c}{P_0}\right).$$

The Nash equilibrium is, therefore,

$$\begin{aligned} q_1^* &= \frac{Q_0}{2}\left(1 - \frac{c}{P_0}\right) \\ q_2^* &= \hat{q}_2(q_1^*) \\ &= \frac{Q_0}{4}\left(1 - \frac{c}{P_0}\right). \end{aligned}$$

It is interesting to note that although Firm 2 has more information than Firm 1 – it knows Firm 1's decision, which has already been made, whereas Firm 1 does not know Firm 2's decision which is still in the future – it is Firm 1 that makes the greater profit (because $q_1^* > q_2^*$).

6.3 The Stackelberg Duopoly Model

Exercise 6.4

Do consumers do better in the Cournot or in the Stackelberg model?

The subgame perfect Nash equilibrium derived above is sometimes called the "Stackelberg equilibrium". However, it is not the only Nash equilibrium in the Stackelberg model. Another Nash equilibrium is for Firm 1 to produce the Cournot quantity and for Firm 2 to produce the Cournot quantity *regardless* of the production of Firm 1. If $q_1 = q_c^*$ then

$$\hat{q}_2 = \frac{Q_0}{2}\left(1 - \frac{q_c^*}{Q_0} - \frac{c}{P_0}\right) = q_c^*$$

So Firm 2's best response to $q_1 = q_c^*$ is $\hat{q}_2 = q_c^*$. If Firm 2 *always* chooses $q_2 = q_c^*$ then Firm 1's profit is

$$\pi_1(q_1, q_c^*) = q_1 \left[P_0 \left(1 - \frac{q_1 + q_c^*}{Q_0}\right) - c \right]$$

The best response (for Firm 1) to $q_2 = q_c^*$ is found from

$$\frac{\partial \pi_1}{\partial q_1}(\hat{q}_1, q_c^*) = 0$$

which gives

$$\begin{aligned}\hat{q}_1 &= \frac{Q_0}{2}\left(1 - \frac{q_c^*}{Q_0} - \frac{c}{P_0}\right) \\ &= q_c^* .\end{aligned}$$

So Firm 1's best response to $q_2 = q_c^*$ is $\hat{q}_1 = q_c^*$. Because, for both firms, the best response to the other firm producing quantity q_c^* is to produce the quantity q_c^*. the pair of strategies (q_c^*, q_c^*) is a Nash equilibrium. Although $q_2 = q_c^*$ is a best response to $q_1 = q_c^*$, it is not a best response to $q_1 \neq q_c^*$. Consequently the Nash equilibrium (q_c^*, q_c^*) is not subgame perfect.

Exercise 6.5

Suppose a firm (the "Entrant") is considering diversifying into a market that is currently monopolised by another firm (the "Incumbent"). Assuming that the market price for the product is given by

$$P(Q) = P_0 \left(1 - \frac{Q}{Q_0}\right)$$

where $Q = q_I + q_E$ the cost of a unit of production for each firm is c and the cost to the Entrant of building manufacturing facilities is C_E, should the Entrant diversify? If the Entrant does diversify, should the incumbent reveal its production plans or keep them a secret?

6.4 War of Attrition

In a *War of Attrition*, two players compete for a resource of value v. This could be two animals competing for ownership of a breeding territory or two supermarkets engaged in a price war. The strategy for each player is a choice of a persistence time, t_i. The model makes three assumptions:

1. The cost of the contest is related only to its duration. There are no other costs (e.g., risk of injury).

2. The player that persists the longest gets all of the resource. If both players quit at the same time, then neither gets the resource.

3. The cost paid by each player is proportional to the shortest persistence time chosen. (That is, no costs are incurred after one player quits and the contest ends.)

Under these assumptions, the payoffs for the two players are

$$\pi_1(t_1, t_2) = \begin{cases} v - ct_2 & \text{if } t_1 > t_2 \\ -ct_1 & \text{if } t_1 \leq t_2 \end{cases}$$

and

$$\pi_2(t_1, t_2) = \begin{cases} v - ct_1 & \text{if } t_2 > t_1 \\ -ct_2 & \text{if } t_2 \leq t_1. \end{cases}$$

There are two pure strategy Nash equilibria. The first is

$$t_1^* = v/c \quad \text{and} \quad t_2^* = 0$$

giving $\pi_1(v/c, 0) = v$ and $\pi_2(v/c, 0) = 0$. This is a Nash equilibrium because, for player 1,

$$\pi_1(t_1, 0) = v \quad \forall t_1 > 0$$
$$\pi_1(0, 0) = 0$$

which gives

$$\pi_1(t_1, t_2^*) \leq \pi_1(t_1^*, t_2^*) \quad \forall t_1.$$

For player 2, we have

$$\pi_2(v/c, t_2) = -ct_2 < 0 \quad \forall t_2 \leq v/c$$
$$\pi_2(v/c, t_2) = 0 \quad \forall t_2 > v/c.$$

Hence

$$\pi_2(t_1^*, t_2) \leq \pi_2(t_1^*, t_2^*) \quad \forall t_2.$$

6.4 War of Attrition

The second pure strategy Nash equilibrium is

$$t_1^* = 0 \quad \text{and} \quad t_2^* = v/c$$

giving $\pi_1(0, v/c) = 0$ and $\pi_2(0, v/c) = v$. The conditions showing that this strategy pair is a Nash equilibrium are the same as the conditions for the first with players 1 and 2 swapped over.

To find a mixed-strategy Nash equilibrium, it is convenient to consider strategies based on the costs of the contest, $x \equiv ct_1$ and $y \equiv ct_2$. In terms of costs, the payoffs are

$$\pi_1(x, y) = \begin{cases} v - y & \text{if } x > y \\ -x & \text{if } x \leq y \end{cases}$$

for player 1, and

$$\pi_2(x, y) = \begin{cases} v - x & \text{if } y > x \\ -y & \text{if } y \leq x \end{cases}$$

for player 2. A mixed strategy σ_1 specifies a choice of cost in the range x to $x + dx$ with probability $p(x)dx$; and σ_2 specifies a similar probability density $q(y)$. The expected payoff to player 1 if he chooses a fixed cost x against a mixed strategy σ_2^* is

$$\pi_1(x, \sigma_2^*) = \int_0^x (v - y)q(y)dy + \int_x^\infty (-x)q(y)dy$$

where the first term arises from the probability that player 2 chooses cost $y \leq x$, and the second term from the probability that player 2 chooses $y > x$.

By extension of the Equality of Payoffs Theorem (Theorem 4.27) for randomising strategies, we must have $\pi_1(x, \sigma_2^*) = \text{constant}$. That is, for fixed σ_2^*, $\pi_1(x, \sigma_2^*)$ is independent of x so

$$\frac{\partial \pi_1}{\partial x} = 0 \ .$$

Now

$$\frac{\partial \pi_1}{\partial x} = \frac{d}{dx} \int_0^x (v - y)q(y)dy - \int_x^\infty q(y)dy - x \frac{d}{dx} \int_x^\infty q(y)dy \ .$$

Using the fundamental theorem of calculus, the first term is

$$\frac{d}{dx} \int_0^x (v - y)q(y)dy = (v - x)q(x) \ .$$

Using the fundamental theorem of calculus and the fact that $q(y)$ is a probability density, we have

$$\frac{d}{dx} \int_x^\infty q(y)dy = \frac{d}{dx}\left[1 - \int_0^x q(y)dy\right]$$
$$= -q(x) \ .$$

So, we have

$$\begin{aligned}\frac{\partial \pi_1}{\partial x} &= (v-x)q(x) - \int_x^\infty q(y)dy + xq(x) \\ &= vq(x) - \int_x^\infty q(y)dy \\ &= 0.\end{aligned}$$

From this we can identify $q(y)$ as an exponential probability density, so we put $q(y) = ke^{-ky}$ where k is a normalisation constant. Because

$$\int_x^\infty q(y)dy = k\int_x^\infty e^{-ky}dy = e^{-kx}$$

we have $vke^{-kx} - e^{-kx} = 0$ or $k = \frac{1}{v}$. Hence

$$q(y) = \frac{1}{v}\exp\left(-\frac{y}{v}\right).$$

In other words, the distribution of costs chosen under the mixed strategy is exponential with mean cost v. The same argument for the other player yields the same distribution of costs:

$$p(x) = \frac{1}{v}\exp\left(-\frac{x}{v}\right)$$

(i.e., the equilibrium is symmetric).

Now that we have found a distribution in terms of costs chosen, we can easily find the Nash equilibrium in terms of the distribution of persistence times chosen. Using

$$p(t) = p(x)\frac{dx}{dt}$$

we have

$$p(t) = \frac{c}{v}\exp\left(-\frac{ct}{v}\right).$$

That is the distribution of times chosen is exponential with mean v/c.

Although $p(t)$ is the distribution of persistence times chosen by each player, it is not the distribution of contest durations. This distribution can be found as follows.

$$\begin{aligned}P(\text{duration} \leq t) &= 1 - P(\text{contest is still going at time } t) \\ &= 1 - P(\text{neither player has quit before } t).\end{aligned}$$

Now

$$\begin{aligned}P(\text{Player } i \text{ doesn't quit before } t) &= \int_t^\infty p(\tau)d\tau \\ &= \exp\left(-\frac{ct}{v}\right).\end{aligned}$$

6.4 War of Attrition

Because the players' decisions are independent, we have

$$\begin{aligned} P(\text{duration} \leq t) &= 1 - \exp\left(-\frac{ct}{v}\right)\exp\left(-\frac{ct}{v}\right) \\ &= 1 - \exp\left(-\frac{2ct}{v}\right) \end{aligned}$$

i.e., contest durations are exponentially distributed with mean $v/(2c)$.

Exercise 6.6

Show that the expected payoff when following this strategy is zero.

Exercise 6.7

In this section, we assumed that the cost of a contest was linearly related to its duration. Find the mixed strategy equilibrium for a War of Attrition in which cost $= kt^2$.

7
Infinite Dynamic Games

7.1 Repeated Games

Consider the following two (related) questions. In the Prisoners' Dilemma, uncooperative behaviour was the predicted outcome although cooperative behaviour would lead to greater payoffs for all players if *everyone* was cooperative. Interpreting the Prisoners' Dilemma as a generalised social interaction, we can ask the question: Is external (e.g., governmental) force required in order to sustain cooperation or can such behaviour be induced in a liberal, individually rational way? In the Cournot duopoly, cartels were not stable. However, in many countries, substantial effort is expended in making and enforcing anti-collusion laws. So it seems that, in reality, there is a risk of cartel formation. How can cartels be stable?

A clue to a possible resolution of these problems lies in the response many people have to the original form of the Prisoners' Dilemma: it is the fear of retaliation in the future that prevents each crook from squealing. In societies, individuals often interact many times during their lives, and the effect on the future seems to be an important consideration when any decision is made. In the business arena, firms make production decisions repeatedly rather than just once. So perhaps the cartel can be sustained by making promises or threats about what will be done in the future.

Inspired by these observations, we will now consider situations in which players interact repeatedly. The payoffs obtained by players in the game will depend on past choices, either because they may condition their strategies on the history of the interaction or because past choices have placed them into a different "state".

Let us consider first the case when there is only one state in which a particular, single-decision game is played. This game is often called the *stage game*. After the stage game has been played, the players again find themselves facing the same situation (i.e., the stage game is repeated). Taken one stage at a time, the only sensible overall strategy is for a player to use their Nash equilibrium strategy for the stage game each time it is played. However, if the game is viewed as a whole, the strategy set becomes much richer. Players may condition their behaviour on the past actions of their opponents or make threats about what they will do in the future if the course of the game does not follow a satisfactory path.

We will restrict our consideration to stage games with a discrete and finite strategy set. This may seem like we are ruling out the possibility of discussing the stability of cartels. However, the restriction does not, in fact, prevent us from considering such questions.

Exercise 7.1

Consider the following finite version of the Cournot duopoly model. Marginal costs are the same for both firms, and the market price is determined by $P(Q) = P_0(1 - Q/Q_0)$ where $Q = q_1 + q_2$ (i.e., the sum of the production quantities chosen by the two firms). Each firm has a pure-strategy set $\{M, C\}$, where

M: produce half the monopolist's optimum quantity $\frac{1}{2}q_m^* = \frac{Q_0}{4}(1 - c/P_0)$

C: produce the Cournot equilibrium quantity $q_c^* = \frac{Q_0}{3}(1 - c/P_0)$

Show that this game has the form of a Prisoners' Dilemma.

So, rather than analyse the discrete Cournot game with its complicated payoffs, we will look at a Prisoners' Dilemma game with payoffs:

		P_2	
		C	D
P_1	C	3,3	0,5
	D	5,0	1,1

The basis for our discussion of repeated games will be the "Iterated Prisoners' Dilemma" in which this stage game will be repeated (iterated) some number of times. Initially, we will discuss games with a finite number of repeats. Then we will consider games with an "infinite" number of repeats, which can be interpreted as meaning that the players are uncertain about when the game will end (see Section 3.5).

7.2 The Iterated Prisoners' Dilemma

First, let us suppose that the Prisoners' Dilemma is repeated just once so that there are 2 stages in all. We solve this just like any dynamic game by backward induction. In the final stage, there is no future interaction so the only consequences for any choice of strategy is the payoff to be gained in that stage. Because the best response is to play D regardless of the opponent's strategy, (D, D) is the Nash equilibrium in this subgame giving a contribution of 1 to the total payoff for each player.

Now consider the first stage. Note that this stage on its own is not a subgame – the subgame starting at the beginning of this stage is the whole game. Because the strategies have been fixed for the final stage, payoffs for the subgame can be calculated by adding the payoffs for the Nash equilibrium in the final stage (i.e., 1 to each player) to the payoffs for the first stage to create a payoff table for the entire game. Note that the pure-strategy set for each player in the entire game is $\mathbf{S} = \{CC, DC, CD, DD\}$ but, because we are only interested in a subgame perfect Nash equilibrium, we only need to consider a subset of the payoff table.

		P_2	
		CD	DD
P_1	CD	4, 4	1, 6
	DD	6, 1	2, 2

The Nash equilibrium in this game is (DD, DD). So the subgame perfect Nash equilibrium for the whole game is to play D in both stages. Note that a player cannot induce cooperation in the first stage by promising to cooperate in the second stage because they would not keep their promise and the other player knows this. Nor can they induce cooperation in the first stage by threatening to defect in the second stage, because this is what happens anyway.

The argument from backward induction can easily be extended to any finite number of repeats, leading to the conclusion that the only solution is for both players to play D in every stage. In other words, bilateral cooperation (or a cartel) is not stable in the finitely repeated Prisoners' Dilemma.

Exercise 7.2

Consider the repeated Prisoners' Dilemma game with 2 stages using the full pure-strategy set $\mathbf{S} = \{CC, DC, CD, DD\}$. Show that both players defecting in each stage is the unique Nash equilibrium.

Now let us consider an infinite number of repeats, indexed by $t = 0, 1, 2, \ldots$. If there is no end to the game (or the players don't know when it will end), then there is no last stage to work backwards from. If the length of the game is infinite, then at any stage there is still an infinite number of stages to go. This suggests we should look for a stationary strategy (because all subgames look the same).

Definition 7.1

A *stationary strategy* is one in which the rule for choosing an action is the same in every stage. Note that this does not necessarily mean that the action chosen in each stage will be the same.

Example 7.2

The strategies "Play C in every stage" and "Play D in every stage" are obviously stationary strategies in the Iterated Prisoners' Dilemma. The conditional strategy "Play C if the other player has never played D and play D otherwise" is also stationary.

The payoff for a stationary strategy is the infinite sum of the payoffs achieved in each stage. Suppose that player i receives a payoff $r_i(t)$ in stage t. Then their total payoff is

$$\sum_{t=0}^{\infty} r_i(t) \ .$$

Unfortunately, this straightforward approach leads to a problem. Consider the strategy $s_C =$"Play C in every stage". If both players use this strategy, the total payoff to either player is

$$\begin{aligned} \pi_i(s_C, s_C) &= \sum_{t=0}^{\infty} 3 \\ &= \infty \end{aligned}$$

whereas, if one player uses the strategy $s_D =$"Play D in every stage", then the total payoff to the defector is

$$\begin{aligned} \pi_1(s_D, s_C) &= \pi_2(s_C, s_D) \\ &= \sum_{t=0}^{\infty} 5 \\ &= \infty \end{aligned}$$

So it is impossible to decide (by comparing total payoffs) whether s_D is better than s_C.

7.2 The Iterated Prisoners' Dilemma

One solution is to discount future payoffs by a factor δ with $0 < \delta < 1$ so that a player's total payoff is

$$\sum_{t=0}^{\infty} \delta^t r_i(t) \,.$$

Depending on the situation being modelled, the discount factor δ represents inflation, uncertainty about whether the game will continue, or a combination of these.

Example 7.3

With the introduction of a discount factor, the payoff if both players always cooperate is

$$\begin{aligned} \pi_i(s_C, s_C) &= \sum_{t=0}^{\infty} 3\delta^t \\ &= \frac{3}{1-\delta} \end{aligned}$$

and the payoff to a unilateral defector is

$$\begin{aligned} \pi_1(s_D, s_C) &= \pi_2(s_C, s_D) \\ &= \sum_{t=0}^{\infty} 1\delta^t \\ &= \frac{5}{1-\delta} \,. \end{aligned}$$

So all payoffs are finite.

Now that we can sensibly compare payoffs achieved by different strategies, can permanent cooperation (a cartel) be stable outcome of the infinitely repeated Prisoners' Dilemma? An answer to this question is provided by the following example.

Example 7.4

Consider the trigger strategy[1] $s_G = $ "Start by cooperating and continue to cooperate until the other player defects, then defect forever after" (this strategy is sometimes given the name *Grim*). If both players adopt this strategy, then

---
[1] This strategy is called a *trigger strategy* because a change in behaviour is triggered by a single defection.

we would observe permanent cooperation and each player would achieve a total payoff

$$\pi_i(s_G, s_G) = 3 + 3\delta + 3\delta^2 + \cdots$$
$$= \frac{3}{1-\delta}.$$

Is (s_G, s_G) a Nash equilibrium?

For simplicity, let us assume that both players are restricted to a pure-strategy set $\mathbf{S} = \{s_G, s_C, s_D\}$ (we will relax this constraint in the next section). Suppose player 1 decides to use the strategy s_C ("always cooperate") instead. Once again, we would observe permanent cooperation and the payoff to each player would be

$$\pi_1(s_C, s_G) = \pi_2(s_C, s_G)$$
$$= \frac{3}{1-\delta}.$$

The same result applies if player 2 decides to switch instead, so neither player can do better (against s_G) by switching to s_C. Now consider player 1 using the alternative strategy s_D ("always defect") against an opponent who uses the trigger strategy s_G. Then the sequence of actions used by the players is as follows:

$t =$	0	1	2	3	4	5	...
Player 1 (s_D):	D	D	D	D	D	D	...
Player 2 (s_G):	C	D	D	D	D	D	...

The payoff for player 1 is

$$\pi_1(s_D, s_G) = 5 + \delta + \delta^2 + \delta^3 + \cdots$$
$$= 5 + \frac{\delta}{1-\delta}.$$

Player 1 cannot do better by switching to s_D from s_G if

$$\frac{3}{1-\delta} \geq 5 + \frac{\delta}{1-\delta}.$$

(The same inequality arises if it is player 2 that switches.) This inequality is satisfied if

$$\delta \geq \frac{1}{2}.$$

So the pair of strategies (s_G, s_G) is a Nash equilibrium if the discounting factor is high enough.

Exercise 7.3

Consider the Iterated Prisoners' Dilemma with pure strategy sets $\mathbf{S}_1 = \mathbf{S}_2 = \{s_D, s_C, s_T, s_A\}$. The strategy s_T is the famous *Tit-For-Tat* ("Begin by cooperating; then do whatever the other player did in the previous stage"), and s_A is a cautious version of Tit-for-Tat with which a player begins by defecting and then does whatever the other player did in the previous stage. What condition does the discount factor have to satisfy in order for (s_T, s_T) to be a Nash equilibrium?

Exercise 7.4

Consider the Iterated Prisoners' Dilemma with pure-strategy sets $\mathbf{S}_1 = \mathbf{S}_2 = \{s_D, s_C, s_G\}$ (i.e., unconditional defection, unconditional cooperation, and the conditional cooperation strategy "Grim"). Write down the strategic form of the game and find all the Nash equilibria.

Exercise 7.5

Consider a game in which the stage game with the payoff table is given below is repeated an infinite number of times and payoffs are discounted by a factor δ ($0 < \delta < 1$) that is common to both players.

	A	B
A	1, 2	3, 1
B	0, 5	2, 3

Assume that the players are limited to selecting pure strategies from the following 3 options.

s_A: Play A in every stage game.

s_B: Play B in every stage game.

s_C: Begin by playing B and continue to play B until your opponent plays A. Once your opponent has played A, play A forever afterwards.

Find the condition on δ such that (s_C, s_C) is a Nash equilibrium.

7.3 Subgame Perfection

The Nash equilibrium where both players adopt the trigger strategy s_G is not a subgame perfect Nash equilibrium for the following reason. At any point in the game, the future of the game (i.e., a subgame) is formally equivalent to the entire game. The possible subgames can be divided into 4 classes: (i) neither player has played D; (ii) both players have played D; (iii) player 1 used D in the last stage but player 2 did not; and (iv) player 2 used D in the last stage but player 1 did not. What does the Nash equilibrium strategy pair (s_G, s_G) specify as the strategies to be used in each of these subgame classes?

In classes (i) and (ii) there is no conflict with the concept of subgame perfection. In class (i), neither player's opponent has played D so the strategy s_G specifies that cooperation should continue until the other player defects (i.e., s_G again). That is the strategy pair specified for class (i) subgames is (s_G, s_G), which is a Nash equilibrium of the subgame because it is a Nash equilibrium of the entire game. In class (ii), both player's opponents have defected so the Nash equilibrium strategy pair (s_G, s_G) specifies that each player should play D forever. That is, the strategy pair adopted in this class of subgame is (s_D, s_D) which is a Nash equilibrium of the subgame since it is a Nash equilibrium of the entire game.

However, in class (iii), s_G specifies that player 1 should switch to using D forever because his opponent has just played D. However, player 1 has not *yet* played D so player 2 should continue to use s_G (which, indeed, results in the use of D from the next round onwards). Thus the Nash equilibrium for the whole game specifies that the strategy pair (s_D, s_G) should be adopted in subgames of class (iii). However, this pair is not a Nash equilibrium for the subgame because player 2 could obtain a greater payoff by using s_D rather than s_G. A similar argument applies to class (iv) subgames. Hence the Nash equilibrium for the entire game does not specify that players play a Nash equilibrium in every possible subgame, hence the Nash equilibrium (s_G, s_G) is not subgame perfect.

Although, (s_G, s_G) is not a subgame perfect Nash equilibrium, a very similar strategy does lead to a subgame perfect Nash equilibrium when it is adopted by both players. Let s_g = "Start by cooperating and continue to cooperate until *either* player defects, then defect forever after". The pair (s_g, s_g) is a subgame perfect Nash equilibrium because it specifies that the players should play (s_D, s_D) in the subgames of classes (iii) and (iv).

Exercise 7.6

Consider the Iterated Prisoners' Dilemma with our usual set of payoffs. Show that both players using the strategy Tit-for-Tat ("Begin by cooperating; then do whatever the other player did in the last stage") is not a subgame perfect Nash equilibrium if the discount factor is $\delta > \frac{2}{3}$.

So far we have allowed repeated games to have only a limited set of strategies. Is it possible to allow more general strategies? If we have found a Nash

7.3 Subgame Perfection

equilibrium candidate from the limited strategy set, can we determine whether any other strategy will do better? For example, in the Iterated Prisoners' Dilemma is the strategy s_G still a Nash equilibrium strategy if more strategies are allowed? If we restrict ourselves to subgame perfect Nash equilibria, then an answer to these questions is provided by the "one-stage deviation principle" for repeated games.

Definition 7.5

A pair of strategies (σ_1, σ_2) satisfies the *one-stage deviation condition* if neither player can increase their payoff by deviating (unilaterally) from their strategy in any single stage and returning to the specified strategy thereafter.[2]

Example 7.6

Consider the Iterated Prisoners' Dilemma and the subgame perfect Nash equilibrium (s_g, s_g) with s_g being the strategy "Start by cooperating and continue to cooperate until either player defects, then defect forever after". Does this pair of strategies satisfy the one-stage deviation condition?

At any given stage, the game will be in one of two classes of subgame: either both players have always cooperated or at least one player has defected in a previous round. If both players have always cooperated, then s_g specifies cooperation in this stage. If either player changes to action D in this stage, then s_g specifies using D forever after. The expected future payoff for the player making this change is

$$5 + \frac{\delta}{1-\delta}$$

which is less than the payoff for continued cooperation if $\delta > \frac{1}{2}$ (which is just the condition for (s_g, s_g) to be a Nash equilibrium). If either player has defected in the past, then s_g specifies defection in this stage. If either player changes to action C in this stage, then s_g still specifies using D forever after. The expected future payoff for the player if they make this change is

$$0 + \frac{\delta}{1-\delta}$$

which is less than the payoff for following the behaviour specified s_g provided $\delta < 1$. Thus the pair (s_g, s_g) satisfies the one-stage deviation condition provided $\frac{1}{2} < \delta < 1$.

---

[2] Compare this with the policy improvement algorithm for Markov decision processes in Section 3.7.

Theorem 7.7 (One-stage Deviation Principle)

A pair of strategies is a subgame perfect Nash equilibrium for a discounted repeated game if and only if it satisfies the one-stage deviation condition.

Proof

For finitely repeated games, the equivalence of subgame perfection and the one-stage deviation condition is guaranteed by the backward induction method. (In fact, this shows that a subgame perfect Nash equilibrium payoff cannot be improved by deviating in any finite number of stages.) Because the definition of subgame perfection implies the one-stage deviation condition for both finitely and infinitely repeated games, it only remains to prove that the one-stage deviation condition implies that a pair constitutes a subgame perfect Nash equilibrium in an infinitely repeated game.

Suppose that, contrary to the statement of the theorem, a strategy pair (σ_1, σ_2) satisfies the one-stage deviation condition but is not a subgame perfect Nash equilibrium. It follows that there is some stage t at which it would be better for one of the players, say player 1, to adopt a different strategy $\hat{\sigma}_1$. That is, there is an ε such that

$$\pi_1(\hat{\sigma}_1, \sigma_2) - \pi_1(\sigma_1, \sigma_2) > 2\varepsilon.$$

Now consider a strategy σ_1' that is the same as $\hat{\sigma}_1$ from stage t up to stage T and is the same as σ_1 from stage T onwards. Because future payoffs are discounted

$$|\pi_1(\hat{\sigma}_1, \sigma_2) - \pi_1(\sigma_1', \sigma_2)| \propto \delta^{T-t}$$

so we can choose a T such that

$$\pi_1(\hat{\sigma}_1, \sigma_2) - \pi_1(\sigma_1', \sigma_2) < \varepsilon.$$

Combining the two inequalities, we get

$$\pi_1(\sigma_1', \sigma_2) - \pi_1(\sigma_1, \sigma_2) > \varepsilon.$$

But σ_1 and σ_1' differ at only a finite number of stages, so this inequality contradicts the one-stage deviation principle for finitely repeated games. It follows that a strategy pair cannot satisfy the one-stage deviation condition without also being a subgame perfect Nash equilibrium. □

Exercise 7.7

Consider an Iterated Prisoners' Dilemma with the following payoffs for the stage game.

	P_2	
	C	D
P_1 C	4,4	0,5
P_1 D	5,0	1,1

Let s_P be the strategy (sometimes called *Pavlov*) "defect if only one player defected in the previous stage (regardless of which player it was); cooperate if either both players cooperated or both players defected in the previous stage". Use the one-stage deviation principle to find a condition for (s_P, s_P) to be a subgame perfect Nash equilibrium.

7.4 Folk Theorems

The *Folk Theorem* was given that name because the result was widely known long before anyone published a formal proof. Since the original result, there have been many variants each of which proves a slightly different result based on slightly different assumptions. However, the general flavour of the result is always the same: if the Nash equilibrium in a static game is socially sub-optimal, players can always do better if the game is repeated and the discount factor is high enough. These theorems are often also called "folk theorems" despite having a well-attested origin. We have just seen an example of a folk theorem in action in the previous section. In the Prisoners' Dilemma, the Nash equilibrium gives each player a poor payoff of 1 compared to the socially optimal payoff of 3. This higher payoff can be achieved (in each stage) by both players as an equilibrium of the repeated game if the discount factor is large enough.

In order to be a bit more specific, we will consider a folk theorem that was proved by Friedman in 1971. To do this, we need the following definitions.

Definition 7.8

Feasible payoff pairs are pairs of payoffs that can be generated by strategies available to the players.

Definition 7.9

Suppose we have a repeated game with discount factor δ. If we interpret δ as the probability that the game continues, then the expected number of stages in which the game is played is

$$T = \frac{1}{1-\delta}$$

Suppose the two players adopt (not necessarily Nash equilibrium) strategies σ_1 and σ_2, then the total expected payoff to player i is $\pi_i(\sigma_1, \sigma_2)$ and the *average payoffs* (per stage) are given by

$$\frac{1}{T}\pi_i(\sigma_1, \sigma_2) = (1-\delta)\pi_i(\sigma_1, \sigma_2).$$

Remark 7.10

The range of feasible payoff pairs in a static game and the range of feasible average payoff pairs if that game is repeated are the same.

Definition 7.11

Individually rational payoff pairs are those average payoffs that exceed the stage game Nash equilibrium payoff for both players.

Example 7.12

In the static Prisoners' Dilemma, pairs of payoffs (π_1, π_2) equal to $(1,1)$, $(0,5)$, $(5,0)$, and $(3,3)$ are obviously feasible since they are generated by combinations of pure strategies. However, although each player could get a payoff as low as 0, the payoff pair $(0,0)$ is not feasible since there is no strategy pair which generates those payoffs for the two players. If player 1 and player 2 use strategy C with probabilities p and q, respectively, the payoffs are given by

$$(\pi_1, \pi_2) = (1 - p + 4q - pq, 1 - q + 4p - pq).$$

Feasible payoff pairs are found by letting p and q take all values between 0 and 1. Individually rational payoff pairs are those for which the payoff to each player is not less than the Nash equilibrium payoff of 1. See Figure 7.1.

Theorem 7.13 (Folk Theorem)

Let (π_1^*, π_2^*) be a pair of Nash equilibrium payoffs for a stage game and let (v_1, v_2) be a feasible payoff pair when the stage game is repeated. For every individually rational pair (v_1, v_2) (i.e., a pair such that $v_1 > \pi_1^*$ and $v_2 >$

7.4 Folk Theorems

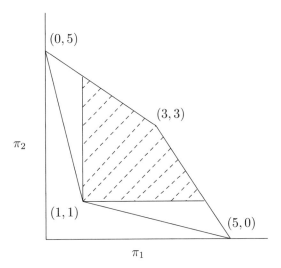

Figure 7.1 Feasible payoffs for the Prisoners' Dilemma lie within the quadrilateral with vertices $(1,1)$, $(0,5)$, $(3,3)$, and $(5,0)$. Feasible average per-stage payoffs for the Iterated Prisoners' Dilemma lie within the same quadrilateral. Individually rational payoffs for the Iterated Prisoners' Dilemma lie in the shaded area.

π_2^*), there exists a $\underline{\delta}$ such that for all $\delta > \underline{\delta}$ there is a subgame perfect Nash equilibrium with payoffs (v_1, v_2).

Proof

Let (σ_1^*, σ_2^*) be the Nash equilibrium that yields the payoff pair (π_1^*, π_2^*). Now suppose that the payoff pair (v_1, v_2) is produced by the players using the actions a_1 and a_2 in every stage (we will consider shortly what happens when this assumption is not valid). Consider the following trigger strategy

> "Begin by agreeing to use action a_i; continue to use a_i as long as both players use the agreed actions; if any player uses an action other than a_i, then use σ_1^* for ever afterwards."

By construction any Nash equilibrium involving these strategies will be subgame perfect, so we only need to find the conditions for a Nash equilibrium. Consider another action a_1' such that the payoff in the stage game for player 1 is $\pi_1(a_1', a_2) > v_1$. Then the total payoff for switching to a_1' against a player using the trigger strategy is not greater than

$$\pi_1(a_1', a_2) + \delta \frac{\pi_1^*}{1-\delta} .$$

It is, therefore, not beneficial to switch to a_1' if $\delta \geq \delta_1$ where

$$\delta_1 = \frac{\pi_1(a_1', a_2) - v_1}{\pi_1(a_1', a_2) - \pi_1^*}.$$

By assumption $\pi_1(a_1', a_2) > v_1 > \pi_1^*$, so $1 > \delta_1 > 0$. A similar argument for player 2 leads to a minimum discount factor δ_2 for player 2. Taking $\underline{\delta} = \max(\delta_1, \delta_2)$ completes this part of the proof.

Now we suppose that the payoffs v_i are achieved by using randomising strategies σ_i. Assume that there exists a randomising device whose output is observed by both players. Assume also that there is an agreed rule for turning the output of the randomising device into a choice of action for each player.[3] These assumptions mean that the strategies themselves (and not just the actions that happen to be taken) are observable. If the strategies are observable in this way, then the previous argument may be repeated with actions a_i and a_i' being replaced by strategies σ_i and σ_i'. □

7.5 Stochastic Games

A stochastic game is defined by a set of states $\mathbf{X}$ with a stage game defined for each state. In each state x, player i can choose actions from a set $\mathbf{A}_i(x)$. One of these stage games is played at each of the discrete times $t = 0, 1, 2, \ldots$. The choice of actions taken by the players in a particular state determines both the immediate rewards obtained by the players and the probability of arriving in any other given state at the next decision point. That is, given that the players are in state x and choose actions $a_1 \in \mathbf{A}_1(x)$ and $a_2 \in \mathbf{A}_2(x)$, the players receive immediate rewards $r_1(x, a_1, a_2)$ and $r_2(x, a_1, a_2)$ and the probability that they find themselves in state x' for the next decision is $p(x'|x, a_1, a_2)$.

Definition 7.14

A strategy is called a *Markov strategy* if the behaviour of a player at time t depends only on the state x. A pure Markov strategy specifies an action $a(x)$ for each state $x \in \mathbf{X}$.

In this section, we will make the following simplifying assumptions.

---

[3] For example, if player 1 has a choice of three actions a, b, and c and is required to choose according to $p(a) = \frac{1}{6}$, $p(b) = \frac{1}{3}$, and $p(c) = \frac{1}{2}$. Then the players may agree that a normal die should be thrown and that player 1 should choose a if the score is 1, b if the score is 2 or 3, and c if the score is 4 or more.

7.5 Stochastic Games

1. The length of the game is not known to the players (i.e., the horizon is infinite).

2. The rewards and transition probabilities are time-independent.

3. The strategies of interest are Markov.

If the number of states is small and the number of actions available in each state is small, then a stochastic game has a simple diagrammatic representation. This diagrammatic representation is best introduced by means of an example. (Compare the description of Markov decision processes in Chapter 3.)

Example 7.15

The set of states is $\mathbf{X} = \{x, z\}$. In state x, both players can choose an action from the sets $\mathbf{A}(x) = \mathbf{A}(y) = \{a, b\}$. The immediate rewards for player 1 for the game in state x are $r_1(x, a, a) = 4$, $r_1(x, a, b) = 5$, $r_1(x, b, a) = 3$, and $r_1(x, b, b) = 2$. This is a zero-sum game so $r_2(x, a_1, a_2) = -r_1(x, a_1, a_2)$ for all action pairs. If players choose the action pair $[a, b]$ in state x, then they move to state z with probability $\frac{1}{2}$ and remain in state x with probability $\frac{1}{2}$. If any other action pair is chosen, the players remain in state x with probability 1. If the players are in state z, then they have the single choice set $\mathbf{A}(z) = \{b\}$ and the immediate rewards $r_1(z, b, b) = r_2(z, b, b) = 0$. Once the players have reached state z, they remain there with probability 1 (so z is a zero-payoff absorbing state). This lengthy description can be presented much more concisely by means of the diagram shown in Figure 7.2.

Consider a game in state x at time t. If we knew the Nash equilibrium strategies for both players from time $t + 1$ onwards, we could calculate the expected future payoffs each player would receive from time $t + 1$ onwards given that they are starting in a particular state. Let us denote the expected future payoff for player i starting in state x by $\pi_i^*(x)$ (with the * indicating that these payoffs are derived using the Nash equilibrium strategies for both players). At time t, the players would then be playing a single-decision game with payoffs given by

$$\pi_i(a_1, a_2) = \left(r_1(x, a_1, a_2) + \delta \sum_{x' \in \mathbf{X}} p(x'|x, a_1, a_2) \pi_1^*(x') \right)$$

where we have assumed that future payoffs are discounted by a factor δ for each time step. We will call this game the *effective game* in state x.

For a Markov strategy, the expected future payoffs in state x are independent of time. Therefore, the payoffs for a Markov-strategy Nash equilibrium

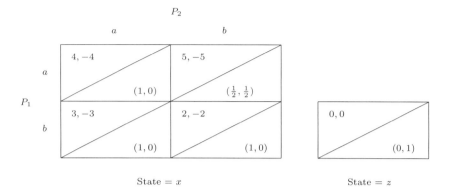

Figure 7.2 The payoffs and state transitions for the stochastic game described in Example 7.15 and solved in Example 7.16 with a discount factor $\delta = \frac{2}{3}$. In state x, the players play a zero-sum game. State z is a zero-payoff absorbing state.

are given by the joint solutions of the following pairs of equations (one for each state $x \in \mathbf{X}$).

$$\pi_1^*(x) = \max_{a_1 \in \mathbf{A}_1(x)} \left(r_1(x, a_1, a_2^*) + \delta \sum_{x' \in \mathbf{X}} p(x'|x, a_1, a_2^*) \pi_1^*(x') \right)$$

$$\pi_2^*(x) = \max_{a_2 \in \mathbf{A}_2(x)} \left(r_2(x, a_1^*, a_2) + \delta \sum_{x' \in \mathbf{X}} p(x'|x, a_1^*, a_2) \pi_2^*(x') \right)$$

Unfortunately, there is no straightforward and infallible method for solving these equations. Nevertheless, a solution can often be found relatively easily, as shown by the following example.

Example 7.16

Consider the stochastic game with the state transitions and payoffs given in Figure 7.2 and discount factor $\delta = \frac{2}{3}$. The value of being in state z is zero for both players. Let v be the present value[4] for player 1 of being in state x (the value for player 2 is $-v$ because this is a zero-sum game). This means that in state x, the players are facing the following effective game.

---
[4] This value is the expected total future payoff.

		P_2	
		a	b
P_1	a	$4+\frac{2}{3}v, -4-\frac{2}{3}v$	$5+\frac{1}{3}v, -5-\frac{1}{3}v$
	b	$3+\frac{2}{3}v, -3-\frac{2}{3}v$	$2+\frac{2}{3}v, -2-\frac{2}{3}v$

Clearly (b,a) is not a Nash equilibrium for any value of v. Ignoring marginal cases, the Nash equilibrium for the effective game in state x will be

1. (a,a) if $v < 3$
2. (b,b) if $v > 9$
3. (a,b) if $3 < v < 9$

Suppose that the players choose (a,a), then $v = 4 + \frac{2}{3}v \Longrightarrow v = 12$, which is inconsistent with the requirement $v < 3$. Now suppose the players choose (b,b), then $v = 2 + \frac{2}{3}v \Longrightarrow v = 6$, which is inconsistent with the requirement $v > 9$. Finally, suppose that the players choose (a,b), then $v = 5 + \frac{1}{3}v \Longrightarrow v = \frac{15}{2}$, which *is* consistent with the requirement $3 < v < 9$. So the unique Markov-strategy Nash equilibrium has the players using the pair of actions (a,b) in state x.

Exercise 7.8

Construct a two-state stochastic game for an Iterated Prisoners' Dilemma problem in which the subgame perfect strategy s_g ("start by cooperating and continue to cooperate until *either* player defects, then defect forever after") can be represented as a Markov strategy. Show that both players using this strategy is a Markov-strategy Nash equilibrium for the stochastic game if $\delta \geq \frac{1}{2}$.

Part III

Evolution

8
Population Games

8.1 Evolutionary Game Theory

So far we have considered two-player games in the framework of Classical Game Theory, where the outcome depends on the choices made by rational and consciously reasoning individuals. The solution for this type of game (the Nash equilibrium) was based on the idea that each player uses a strategy that is a best response to the strategy chosen by the other, so neither would change what they were doing. For symmetric Nash equilibria, (σ^*, σ^*), we can give an alternative interpretation of the Nash equilibrium by placing the game in a population context. In a population where everyone uses strategy σ^*, the best thing to do is follow the crowd; so if the population starts with everyone using σ^*, then it will remain that way – the *population* is in equilibrium. Nash himself introduced this view, calling it the "mass action interpretation". A natural question to ask is then: What happens if the population is close to, but not actually at, its equilibrium configuration? Does the population tend to evolve towards the equilibrium or does it move away? This question can be investigated using Evolutionary Game Theory, which was invented for biological models but has now been adopted by some economists.

Evolutionary Game Theory considers a population of decision makers. In the population, the frequency with which a particular decision is made can change over time in response to the decisions made by all individuals in the population (i.e., the population *evolves*). In the biological interpretation of this evolution, a population consists of animals each of which are genetically programmed to use some strategy that is inherited by its offspring[1]. Initially, the population may consist of animals using different strategies. The payoff to an individual adopting a strategy σ is identified with the fitness (expected number of offspring) for that type in the current population. Animals with higher fitness leave more offspring (by definition) so in the next generation the composition of the population will change. In the economic interpretation, the population changes because people play the game many times and consciously switch strategies. People are likely to switch to those strategies that give better payoffs and away from those that give poor payoffs.

8.2 Evolutionarily Stable Strategies

As with any dynamical system, one interesting question is: What are the end-points (if there are any) of the evolution? One type of evolutionary end-point is called an evolutionarily stable strategy (ESS).

Definition 8.1

Consider an infinite population of individuals that can use some set of pure strategies, **S**. A *population profile* is a vector **x** that gives a probability $x(s)$ with which each strategy $s \in \mathbf{S}$ is played in the population.

A population profile need not correspond to a strategy adopted by any member of the population.

Example 8.2

Consider a population of individuals that can use two strategies s_1 and s_2. If every member of the population randomises by playing each of the two pure strategies with probability $\frac{1}{2}$, then the population profile is $\mathbf{x} = (\frac{1}{2}, \frac{1}{2})$. In this case, the population profile is identical to the mixed strategy adopted by all population members. On the other hand, if half the population adopt the strategy s_1 and the other half adopt the strategy s_2, then the population profile is again $\mathbf{x} = (\frac{1}{2}, \frac{1}{2})$, which is not the same as the strategy adopted by *any* member of the population.

---

[1] This *phenotypic gambit* was discussed in Section 1.4.

8.2 Evolutionarily Stable Strategies

Exercise 8.1

(a) Give three ways in which a population with profile $\mathbf{x} = \left(\frac{5}{12}, \frac{7}{12}\right)$ might arise. (b) Consider an strategy set $\mathbf{S} = \{s_1, s_2, s_3\}$. If the population consists of 40% of individuals using the strategy $(\frac{1}{2}, 0, \frac{1}{2})$ and 60% using $(\frac{1}{4}, \frac{3}{4}, 0)$, what is the population profile?

Consider a particular individual in a population with profile $\mathbf{x}$. If that individual uses a strategy σ, then the payoff to that individual is denoted $\pi(\sigma, \mathbf{x})$. (Note that the other "player" is actually the population and does not have a payoff.) The payoff for this strategy is calculated by

$$\pi(\sigma, \mathbf{x}) = \sum_{s \in S} p(s)\pi(s, \mathbf{x}) \ .$$

These payoffs represent the number of descendants (either through breeding or through imitation) that each type of individual has. Therefore, the payoffs determine the evolution of the population.

Example 8.3

Consider a population of N animals in which individuals are programmed to use one of two strategies s_1 and s_2. Suppose that 50% of the animals use each of the strategies, i.e., $\mathbf{x} = (\frac{1}{2}, \frac{1}{2})$ and that, for this current population profile,

$$\pi(s_1, \mathbf{x}) = 6 \quad \text{and} \quad \pi(s_2, \mathbf{x}) = 4 \ .$$

In the next generation, there will be $6N/2$ individuals using s_1 and $4N/2$ individuals using s_2, so the new population profile will be $\mathbf{x} = (0.6, 0.4)$.

In order to proceed with the next generation we need to determine how the payoffs change when the population profile alters: that is, we need to know how $\pi(s, \mathbf{x})$ behaves as a function of $\mathbf{x}$. Mathematically, the distinction is whether the payoff is a linear or non-linear function of the various probabilities $x(s)$. From a modelling viewpoint, we distinguish between two types of population game: games against the field and pairwise contests.

Definition 8.4

A *game against the field* is one in which there is no specific "opponent" for a given individual – their payoff depends on what everyone in the population is doing.

Games against the field are quite different from the games considered by Classical Game Theory: one consequence of the population-wide interaction is that the payoff to the given individual is not (necessarily) linear in the probabilities $x(s)$ with which the pure strategies are played by population members.

Definition 8.5

A *pairwise contest* game describes a situation in which a given individual plays against an opponent that has been randomly selected (by Nature) from the population and the payoff depends just on what both individuals do.

Pairwise contests are much more like games from Classical Game Theory in that we can write

$$\pi(\sigma, \mathbf{x}) = \sum_{s \in \mathbf{S}} \sum_{s' \in \mathbf{S}} p(s) x(s') \pi(s, s')$$

for suitably defined pairwise payoffs $\pi(s, s')$.

Sometimes games against the field are referred to as "frequency-dependent selection", and the word "game" is reserved for pairwise contests where there is an identifiable interaction between two individuals. However, general population games may include interactions of both types, so we will refer to both of them as "games". This will also help us to maintain the distinction between Classical and Evolutionary Game Theory, which is often obscured when only pairwise contests are considered.

We are interested in the end points of the evolution of the population. In other words, we wish to find the conditions under which the population is stable. Let $\mathbf{x}^*$ be the profile generated by a population of individuals who all adopt strategy σ^* (i.e., $\mathbf{x}^* = \sigma^*$). A necessary condition for evolutionary stability is

$$\sigma^* \in \underset{\sigma \in \boldsymbol{\Sigma}}{\operatorname{argmax}} \, \pi(\sigma, \mathbf{x}^*).$$

So, at an equilibrium, the strategy adopted by individuals must be a best response to the population profile that it generates. Furthermore, we have the population equivalent of Theorem 4.27.

Theorem 8.6

Let σ^* be a strategy that generates a population profile $\mathbf{x}^*$. Let $\mathbf{S}^*$ be the support of σ^*. If the population is stable, then $\pi(s, \mathbf{x}^*) = \pi(\sigma^*, \mathbf{x}^*) \; \forall s \in \mathbf{S}^*$.

8.2 Evolutionarily Stable Strategies

Proof

If the set $\mathbf{S}^*$ contains only one strategy, then the theorem is trivially true. Suppose now that the set $\mathbf{S}^*$ contains more than one strategy. If the theorem is not true, then at least one strategy gives a higher payoff than $\pi(\sigma^*, \mathbf{x}^*)$. Let s' be the action that gives the greatest such payoff. Then

$$\begin{aligned}
\pi(\sigma^*, \mathbf{x}^*) &= \sum_{s \in \mathbf{S}^*} p^*(s) \pi(s, \mathbf{x}^*) \\
&= \sum_{s \neq s'} p^*(s) \pi(s, \mathbf{x}^*) + p^*(s') \pi(s', \mathbf{x}^*) \\
&< \sum_{s \neq s'} p^*(s) \pi(s', \mathbf{x}^*) + p^*(s') \pi(s', \mathbf{x}^*) \\
&= \pi(s', \mathbf{x}^*)
\end{aligned}$$

which contradicts the original assumption that the population is stable. □

If σ^* is a unique best response to $\mathbf{x}^*$, then the evolution of the population clearly stops. However, if there is some other strategy that does equally well in the population with profile $\mathbf{x}^*$, then the population could drift in the direction of the other strategy and its corresponding population profile – unless it is prevented from doing so.

Definition 8.7

Consider a population where (initially) all the individuals adopt some strategy σ^*. Now suppose a (genetic) mutation occurs and a small proportion ε of individuals use some other strategy σ. The new population (i.e., after the appearance of the mutants) is called the *post-entry population* and will be denoted by $\mathbf{x}_\varepsilon$.

Example 8.8

Consider a population in which $\mathbf{S} = \{s_1, s_2\}$ and $\sigma^* = (\frac{1}{2}, \frac{1}{2})$. Suppose the mutant strategy is $\sigma = (\frac{3}{4}, \frac{1}{4})$. Then[2]

$$\begin{aligned}
\mathbf{x}_\varepsilon &= (1 - \varepsilon) \sigma^* + \varepsilon \sigma \\
&= (1 - \varepsilon)(\frac{1}{2}, \frac{1}{2}) + \varepsilon(\frac{3}{4}, \frac{1}{4})
\end{aligned}$$

---

[2] There is a slight abuse of notation here because strategies and population profiles are different objects. What we mean is that the components of the two vectors are equal, i.e., $x(s) = p(s), \forall s \in \mathbf{S}$.

$$= (\frac{1}{2} + \frac{\varepsilon}{4}, \frac{1}{2} - \frac{\varepsilon}{4})$$

Definition 8.9

A mixed strategy σ^* is an *ESS* if there exists an $\bar{\varepsilon}$ such that for every $0 < \varepsilon < \bar{\varepsilon}$ and every $\sigma \neq \sigma^*$

$$\pi(\sigma^*, \mathbf{x}_\varepsilon) > \pi(\sigma, \mathbf{x}_\varepsilon).$$

In other words, a strategy σ^* is an ESS if mutants that adopt any other strategy σ leave fewer offspring in the post-entry population, provided the proportion of mutants is sufficiently small. In the next two sections, we consider the application of this definition – first in a game against the field and then a pairwise contest.

8.3 Games Against the Field

Have you ever wondered why the ratio of males to females in (most) human (and other animal) populations is 50:50? One way of phrasing the answer is because that ratio is an ESS.

Example 8.10

Consider game defined by the following conditions.

1. The proportion of males in the population is μ and the proportion of females is $1 - \mu$.

2. Each female mates once and produces n offspring.

3. Males mate $(1 - \mu)/\mu$ times, on average.

4. Only females "make decisions"[3].

For simplicity, assume that the females' available pure strategies are either to produce no female offspring (s_1) or to produce no male offspring (s_2). With this strategy set, a general strategy $\sigma = (p, 1 - p)$ produces a proportion p of male offspring. A population profile $\mathbf{x} = (x, 1 - x)$ produces a sex ratio $\mu = x$, so we can write the population profile naturally in terms of the sex ratio as $\mathbf{x} = (\mu, 1 - \mu)$.

---
[3] That is, only female genes affect the sex ratio of offspring, so Natural Selection acts only on females.

Because the number of offspring is fixed at n, that clearly cannot be used as the payoff (fitness) for a strategy. However, the number of grandchildren does vary, so we will use that as the payoff. So, in a population with profile $\mathbf{x} = (\mu, 1 - \mu)$, the payoffs are

$$\pi(s_1, \mathbf{x}) = n^2 \frac{1-\mu}{\mu} \qquad (8.1)$$

$$\pi(s_2, \mathbf{x}) = n^2 \qquad (8.2)$$

(n female children each produce n grandchildren for the female, and n male children each get $(1 - \mu)/\mu$ matings and produce n grandchildren from each mating). The fitness of a mixed strategy $\sigma = (p, 1 - p)$ is, therefore,

$$\pi(\sigma, \mathbf{x}) = n^2 \left((1-p) + p \left(\frac{1-\mu}{\mu} \right) \right).$$

Because n is independent of the strategy chosen, we can set $n = 1$ for ease of calculation (we are, after all, interested in the sex *ratio*).

At this point, it might be tempting to construct a payoff table for the game, such as

		Population	
		$x = 1$	$x = 0$
Female	s_1	$\pi(s_1, x = 1)$	$\pi(s_1, x = 0)$
	s_2	$\pi(s_2, x = 1)$	$\pi(s_2, x = 0)$

or, even

		Population	
		s_1	s_2
Female	s_1	$\pi(s_1, s_1)$	$\pi(s_1, s_2)$
	s_2	$\pi(s_2, s_1)$	$\pi(s_2, s_2)$

However, we should not do this for two reasons. First, the profile $(\mathbf{x}_0 = (0, 1))$ leads to $\mu = 0$, which means the payoff for s_1 is undefined. Second, it might tempt us to believe that the pure-strategy payoffs in a general population are

$$\pi(s_i, \mathbf{x}) = x\pi(s_i, s_1) + (1 - x)\pi(s_i, s_2)$$

which they are not: in Equation 8.1, the payoff to the strategy s_1 is a non-linear function of the population profile.

The first of these problems is an affliction of the simple way we have set up the basic model. However, because the evolutionarily stable population will turn out to be well away from this state, we can ignore this problem and continue to use the simple formulation. An alternative approach is to specify pure strategies that always produce a non-zero proportion of males – see Exercise 8.3.

Let us now try to find an ESS. Consider the following three cases.

1. If $\mu < \frac{1}{2}$, then females using s_1 (all male offspring) have more grandchildren, which (eventually) causes μ to rise. So, s_1 is not an ESS.

2. If $\mu > \frac{1}{2}$, then females using s_2 (all female offspring) have more grandchildren, which causes μ to fall. So, s_2 is not an ESS.

3. $\sigma^* = (\frac{1}{2}, \frac{1}{2})$ is a potential ESS, because by Theorem 8.6

$$\pi(s_1, \mathbf{x}^*) = \pi(s_2, \mathbf{x}^*) = \pi(\sigma^*, \mathbf{x}^*) \tag{8.3}$$

if the population profile is $\mathbf{x}^* = (\frac{1}{2}, \frac{1}{2})$ (i.e. $\mu = \frac{1}{2}$).

Because Equation 8.3 is a necessary but not sufficient condition for evolutionary stability, we need to check that $\sigma^* = (\frac{1}{2}, \frac{1}{2})$ is, in fact, an ESS. Let $\sigma = (p, 1-p)$ then

$$\mathbf{x}_\varepsilon = (1-\varepsilon)\sigma^* + \varepsilon\sigma$$

and

$$\mu_\varepsilon = \frac{1}{2}(1-\varepsilon) + \varepsilon p = \frac{1}{2} + \varepsilon\left(p - \frac{1}{2}\right).$$

The ESS condition is

$$\pi(\sigma^*, \mathbf{x}_\varepsilon) > \pi(\sigma, \mathbf{x}_\varepsilon)$$

where

$$\pi(\sigma^*, \mathbf{x}_\varepsilon) = \frac{1}{2} + \frac{1}{2}\left(\frac{1-\mu_\varepsilon}{\mu_\varepsilon}\right)$$

and

$$\pi(\sigma, \mathbf{x}_\varepsilon) = (1-p) + p\left(\frac{1-\mu_\varepsilon}{\mu_\varepsilon}\right).$$

The difference between the payoffs is

$$\begin{aligned}
\pi(\sigma^*, \mathbf{x}_\varepsilon) - \pi(\sigma, \mathbf{x}_\varepsilon) &= \left(p - \frac{1}{2}\right) + \left(\frac{1}{2} - p\right)\left(\frac{1-\mu_\varepsilon}{\mu_\varepsilon}\right) \\
&= \left(\frac{1}{2} - p\right)\left[\frac{1-\mu_\varepsilon}{\mu_\varepsilon} - 1\right] \\
&= \left(\frac{1}{2} - p\right)\left[\frac{1-2\mu_\varepsilon}{\mu_\varepsilon}\right]
\end{aligned}$$

If this difference is positive for *any* $\sigma = (p, 1-p)$ with $p \neq \frac{1}{2}$ then σ^* is an ESS. Because

$$\begin{aligned}
p < \frac{1}{2} &\implies \mu_\varepsilon < \frac{1}{2} \\
&\implies \pi(\sigma^*, \mathbf{x}_\varepsilon) > \pi(\sigma, \mathbf{x}_\varepsilon)
\end{aligned}$$

and

$$p > \frac{1}{2} \implies \mu_\varepsilon > \frac{1}{2}$$
$$\implies \pi(\sigma^*, \mathbf{x}_\varepsilon) > \pi(\sigma, \mathbf{x}_\varepsilon)$$

the mixed strategy $\sigma^* = (\frac{1}{2}, \frac{1}{2})$ is an ESS.

Note that we have shown that a monomorphic ("one form") population where everyone uses the strategy σ^* is evolutionarily stable, with profile $\mathbf{x}^*$. However, in this population individuals using s_1 and individuals using s_2 have the same fitness as individuals using σ^*. So is a polymorphic ("many form") population in which, for example, 50% of animals use s_1 and 50% use s_2 also stable? This polymorphic population also generates a profile $\mathbf{x}^*$, but neither of these strategies is an ESS on its own and the ESS formalism cannot deal with polymorphisms – we will address this question in the next chapter.

Exercise 8.2

Consider a simplified version of the Internet. There are two operating systems available to computer users: L and W. A user of system W has a basic utility of 1, but L is a better operating system so a user of L has a basic utility of 2. If two computers have the same operating system, then they can communicate over the network. (N.B. this is not a necessary requirement on the real Internet.) A user's utility rises linearly with the proportion of computers that can be communicated with, up to a maximum increment of 2. Let x be the proportion of W-users, then $\pi(W, x) = 1 + 2x$ and $\pi(L, x) = 2 + 2(1 - x)$. What are the ESSs in this population game?

Exercise 8.3

Consider a sex ratio game in which females can choose between two pure strategies:

s_1: produce n offspring in which the proportion of males is 0.8

s_2: produce n offspring in which the proportion of males is 0.2

Consider a female using the mixed strategy $\sigma = (p, 1 - p)$ in a population with a proportion of males $= \mu$. (a) Find the expected number of grandchildren for this female. (b) Hence show that, in a monomorphic population, the only possible evolutionarily stable sex ratio has $\mu = \frac{1}{2}$. (c) Find the strategy which leads to $\mu = \frac{1}{2}$ in a monomorphic population and show that it is evolutionarily stable.

8.4 Pairwise Contest Games

As an example of a pairwise contest, let's look at one of the first evolutionary games that was invented to model conflict between animals. The two basic types are called "Hawks" and "Doves". However, they are not intended to represent two different species of animal; instead they represent two types of behaviour (i.e., actions or pure strategies) that could be exhibited by animals of the same species. The terminology arises from human behaviour where those who advocate pre-emptive military solutions to international problems are called "Hawks" while those who would prefer a more diplomatic approach are called "Doves".

The biological significance of the Hawk-Dove game is that it provides an alternative to group-selectionist arguments for the persistence of species whose members have potentially lethal attributes (teeth, horns, etc.). The question to be answered is the following. Because it is obviously advantageous to fight for a resource (having it all is better than sharing), why don't animals always end up killing (or at least seriously maiming) each other? The group-selectionist answer is that any species following this strategy would die out pretty quickly, so animals hold back from all out contests "for the good of the species". The "problem" with this is that it seems to require more than just individual-based Natural Selection to be driving Evolution. So, if group selection is the *only* possible answer, then that would be a very important result. However, the Hawk-Dove game shows that there is an alternative – one that is based fairly and squarely on the action of Natural Selection on *individuals*. So, applying Occam's Razor, there is no need to invoke group selection.[4]

Example 8.11 (The Hawk-Dove Game)

Individuals can use one of two possible pure strategies

H : Be aggressive ("be a Hawk")
D : Be non-aggressive ("be a Dove").

In general, an individual can use a randomised strategy which is to be aggressive with probability p, i.e., $\sigma = (p, 1-p)$. A population consists of animals that are aggressive with probability x, i.e., $\mathbf{x} = (x, 1-x)$, which can arise because (i) in a monomorphic population, everyone uses the strategy $\sigma = (x, 1-x)$, or (ii) in

---

[4] William of Occam (1285–1349) was a Franciscan friar. His logical principle, as expressed in *Summa Totius Logicae*, states "frustra fit per plura quod potest fieri per pauciora" (it is pointless to do with more what can be done with less). This approach was echoed later by Isaac Newton (1642–1727) in the *Philosophiae Naturalis Principia Mathematica*: "We are to admit no more causes of natural things than such as are both true and sufficient to explain their appearances".

a polymorphic population a fraction x of the population use $\sigma_H = (1,0)$ and a fraction $1-x$ use $\sigma_D = (0,1)$. We will consider only monomorphic populations for the moment.

At various times, individuals in this population may come into conflict over a resource of value v. This could be food, a breeding site, etc. The outcome of a conflict depends on the types of the two individuals that meet. If a Hawk and a Dove meet, then the Hawk gains the resource without a fight. If two Doves meet, then they "share" the resource. If two Hawks meet, then there is a fight and each individual has an equal chance of winning. The winner gets the resource and the loser pays a cost (e.g., injury) of c. The payoffs for a focal individual are then

$$\pi(\sigma, \mathbf{x}) = px\frac{v-c}{2} + p(1-x)v + (1-p)(1-x)\frac{v}{2} .$$

To make things interesting, we assume $v < c$ (this is then "*the* Hawk-Dove game"). It is easy to see that there is no pure-strategy ESS. In a population of Doves, $x = 0$, and

$$\begin{aligned}\pi(\sigma, \mathbf{x}_D) &= pv + (1-p)\frac{v}{2} \\ &= (1+p)\frac{v}{2}\end{aligned}$$

so the best response to this population is to play Hawk (i.e., individuals using the strategy $\sigma_H = (1,0)$ will do best in this population). As a consequence, the proportion of more aggressive individuals will increase (i.e., x increases). In a population of Hawks, $x = 1$, and

$$\pi(\sigma, \mathbf{x}_H) = p\frac{v-c}{2}$$

so the best response to this population is to play Dove (i.e., $p = 0$ – remember that we have assumed $v < c$).

Is there a mixed-strategy ESS, $\sigma^* = (p^*, 1-p^*)$? For σ^* to be an ESS, it must be a best response to the the population $\mathbf{x}^* = (p^*, 1-p^*)$ that it generates. In the population $\mathbf{x}^*$, the payoff to an arbitrary strategy $\sigma = (p, 1-p)$ is

$$\begin{aligned}\pi(\sigma, \mathbf{x}^*) &= pp^*\frac{v-c}{2} + p(1-p^*)v + (1-p)(1-p^*)\frac{v}{2} \\ &= (1-p^*)\frac{v}{2} + \frac{pc}{2}\left[\frac{v}{c} - p^*\right]\end{aligned}$$

If $p^* < v/c$ then best response is $\hat{p} = 1$ (i.e., $\hat{p} \neq p^*$). If $p^* > v/c$, then the best response is $\hat{p} = 0$ (i.e., $\hat{p} \neq p^*$). If $p^* = v/c$, then any choice of p (*including* p^*) gives the same payoff (i.e., $\pi(\sigma^*, \mathbf{x}^*) = \pi(\sigma, \mathbf{x}^*)$) and is a best response to $\mathbf{x}^*$. So we have

$$\sigma^* = \left(\frac{v}{c}, 1 - \frac{v}{c}\right)$$

(i.e., be aggressive with probability v/c) as a candidate for an ESS. Recall that $v < c$, so this is a proper mixed strategy.

To confirm that σ^* is an ESS we must show that, for every $\sigma \neq \sigma^*$,

$$\pi(\sigma^*, \mathbf{x}_\varepsilon) > \pi(\sigma, \mathbf{x}_\varepsilon)$$

where the post-entry population profile is

$$\begin{aligned}\mathbf{x}_\varepsilon &= ((1-\varepsilon)p^* + \varepsilon p, ((1-\varepsilon)(1-p^*) + \varepsilon(1-p)) \\ &= (p^* + \varepsilon(p - p^*), 1 - p^* + \varepsilon(p^* - p)) \,.\end{aligned}$$

Now

$$\pi(\sigma^*, \mathbf{x}_\varepsilon) = p^*(p^* + \varepsilon(p - p^*))\frac{v-c}{2} + p^*(1 - p^* + \varepsilon(p^* - p))v$$
$$+ (1 - p^*)(1 - p^* + \varepsilon(p^* - p))\frac{v}{2}$$

and

$$\pi(\sigma, \mathbf{x}_\varepsilon) = p(p^* + \varepsilon(p - p^*))\frac{v-c}{2} + p(1 - p^* + \varepsilon(p^* - p))v$$
$$+ (1 - p)(1 - p^* + \varepsilon(p^* - p))\frac{v}{2} \,.$$

So, after a few lines of algebra (using the fact that $p^* = v/c$), we find

$$\begin{aligned}\pi(\sigma^*, \mathbf{x}_\varepsilon) - \pi(\sigma, \mathbf{x}_\varepsilon) &= \frac{\varepsilon c}{2}(p^* - p)^2 \\ &> 0 \qquad \forall p \neq p^* \quad \text{(i.e. } \forall \sigma \neq \sigma^*\text{)}\end{aligned}$$

which proves that σ^* is an ESS.

Exercise 8.4

Consider a Hawk-Dove game with $v \geq c$. Show that playing H is an ESS.

In order to provide a change of emphasis, we will now consider an economic model for the introduction of currency as a medium of exchange. Because we do not want to get mired in the economic details, the model will be rather schematic.

Example 8.12 (The Evolution of Money)

On a remote, tropical island the inhabitants realise that trade could be conducted more efficiently if they used something as a token for buying and selling, rather than exchanging goods directly. On the island there are two objects that could be used for this purpose: beads and shells. Each individual can choose

8.4 Pairwise Contest Games

to use beads or to use shells, but he can only complete a transaction if the person he is attempting to trade with uses the same token. For simplicity, we will normalise the payoffs so that a trader gets a utility increment of 1 if the transaction succeeds and 0 if it fails.

The general strategy available to an individual is to use beads with probability p: i.e., $\sigma = (p, 1-p)$. A general population profile is $\mathbf{x} = (x, 1-x)$: i.e., the proportion of individuals in the population who are using beads is x. Assuming that an individual attempts to trade with a randomly selected member of the population, his payoff is

$$\begin{aligned} \pi(\sigma, \mathbf{x}) &= px + (1-p)(1-x) \\ &= (1-x) + p(2x-1) \, . \end{aligned}$$

From this we see that

$$x > \frac{1}{2} \implies \hat{p} = 1 \quad \text{and} \quad p = 1 \implies x = 1$$

so $\sigma_b^* = (1, 0)$ is a potential ESS, with a corresponding population profile $\mathbf{x} = (1, 0)$. The post-entry population is

$$\begin{aligned} \mathbf{x}_\varepsilon &= (1-\varepsilon)(1,0) + \varepsilon(p, 1-p) \\ &= (1 - \varepsilon(1-p), \varepsilon(1-p)) \end{aligned}$$

In this population, the payoff for an arbitrary strategy is

$$\pi(\sigma, \mathbf{x}_\varepsilon) = \varepsilon(1-p) + p(1 - 2\varepsilon(1-p))$$

and the payoff for the candidate ESS is

$$\pi(\sigma_b^*, \mathbf{x}_\varepsilon) = 1 - \varepsilon(1-p) \, .$$

So

$$\begin{aligned} \pi(\sigma_b^*, \mathbf{x}_\varepsilon) - \pi(\sigma, \mathbf{x}_\varepsilon) &> 0 \\ \iff (1-p)(1 - 2\varepsilon(1-p)) &> 0 \, . \end{aligned}$$

Now, $\forall p \neq p^*$ we have $1 - p > 0$, so σ_b^* is an ESS if and only if $\varepsilon(1-p) < \frac{1}{2}$. That is, the $\bar{\varepsilon}$ mentioned in definition 8.9 of an ESS is equal to a half.

The strategy $\sigma_s^* = (0, 1)$ is another ESS because in the relevant post-entry population, $\mathbf{x}_\varepsilon = (\varepsilon p, 1 - \varepsilon p)$, the payoff for an arbitrary strategy is

$$\pi(\sigma, \mathbf{x}_\varepsilon) = (1 - \varepsilon p) - p(1 - 2\varepsilon p)$$

and the payoff for the candidate ESS is

$$\pi(\sigma_b^*, \mathbf{x}_\varepsilon) = 1 - \varepsilon p \, .$$

So
$$\pi(\sigma_s^*, \mathbf{x}_\varepsilon) - \pi(\sigma, \mathbf{x}_\varepsilon) > 0$$
$$\iff p(1 - 2\varepsilon p) > 0.$$

Now, $\forall p \neq p^*$ we have $p > 0$, so σ_b^* is an ESS if and only if $\varepsilon p < \frac{1}{2}$ – i.e., $\bar{\varepsilon} = \frac{1}{2}$. The final candidate for an ESS is $\sigma_m^* = \left(\frac{1}{2}, \frac{1}{2}\right)$, because

$$x = \frac{1}{2} \implies \hat{p} \in [0, 1] \implies x \in [0, 1]$$

(including, of course, $x = \frac{1}{2}$). Consider the post-entry population

$$\begin{aligned}
\mathbf{x}_\varepsilon &= (1 - \varepsilon)\left(\frac{1}{2}, \frac{1}{2}\right) + \varepsilon(p, 1 - p) \\
&= \left(\frac{1}{2} - \frac{1}{2}\varepsilon(1 - 2p), \frac{1}{2} + \frac{1}{2}\varepsilon(1 - 2p)\right)
\end{aligned}$$

The payoff for an arbitrary strategy is $\pi(\sigma, \mathbf{x}_\varepsilon) = \frac{1}{2} + \frac{1}{2}\varepsilon(1-2p)^2$ and the payoff for the candidate ESS is $\pi(\sigma_m^*, \mathbf{x}_\varepsilon) = \frac{1}{2}$. So

$$\pi(\sigma_m^*, \mathbf{x}_\varepsilon) - \pi(\sigma, \mathbf{x}_\varepsilon) > 0$$
$$\iff -\frac{1}{2}\varepsilon(1 - 2p)^2 > 0.$$

Because $\varepsilon > 0$ and $p \neq \frac{1}{2}$, this condition cannot be satisfied; so σ_m^* is not an ESS.

Putting the three results together, we can see that the population of islanders will evolve to use either beads or shells as currency; the final outcome depends on the proportion of islanders that initially chooses beads.

Exercise 8.5

Consider a Prisoners' Dilemma where the payoffs for an interaction between two individuals are given by

		P_2	
		C	D
P_1	C	3,3	0,5
	D	5,0	1,1

If a population of individuals play this pairwise contest, what is the ESS?

8.5 ESSs and Nash Equilibria

In this section, we show that the ESSs in a pairwise contest population game correspond to a (possibly empty) subset of the set of Nash equilibria for an associated two-player game. We restrict our attention to pairwise contest games, because the concept of a Nash equilibrium has no meaning for a game against the field.

In a pairwise contest population game, the payoff to a focal individual using σ in a population with profile $\mathbf{x}$ is

$$\pi(\sigma, \mathbf{x}) = \sum_{s \in \mathbf{S}} \sum_{s' \in \mathbf{S}} p(s) x(s') \pi(s, s') \,. \tag{8.4}$$

This payoff is the same as would be achieved in a two-player game against an opponent using a strategy σ' that assigns $p'(s) = x(s) \forall s \in \mathbf{S}$, so we can always associate a two-player game with a population game involving pairwise contests.

Definition 8.13

If a pairwise contest population game has payoffs given by Equation 8.4, then the *associated two-player game* is the game with payoffs given by the numbers[5] $\pi_1(s, s') = \pi(s, s') = \pi_2(s', s)$.

In a monomorphic population, if σ^* is an ESS, then $\mathbf{x}^* = \sigma^*$. So, if there is a Nash equilibrium in the associated game corresponding to the ESS in the population game, then it must be of the form (σ^*, σ^*). That is, a symmetric Nash equilibrium can be associated with an ESS but an asymmetric one cannot.

Theorem 8.14

Let σ^* be an ESS in a pairwise contest then, $\forall \sigma \neq \sigma^*$ either

1. $\pi(\sigma^*, \sigma^*) > \pi(\sigma, \sigma^*)$, or
2. $\pi(\sigma^*, \sigma^*) = \pi(\sigma, \sigma^*)$ and $\pi(\sigma^*, \sigma) > \pi(\sigma, \sigma)$

Conversely, if either (1) or (2) holds for each $\sigma \neq \sigma^*$ in a two-player game, then σ^* is an ESS in the corresponding population game.

---
[5] By convention, player 1 is taken to be the focal player in the population game.

Proof

If σ^* is an ESS, then, by definition (for ε sufficiently small),

$$\pi(\sigma^*, \mathbf{x}_\varepsilon) > \pi(\sigma, \mathbf{x}_\varepsilon)$$

where $\mathbf{x}_\varepsilon = (1-\varepsilon)\sigma^* + \varepsilon\sigma$. For pairwise contests, this condition can be rewritten as

$$(1-\varepsilon)\pi(\sigma^*, \sigma^*) + \varepsilon\pi(\sigma^*, \sigma) > (1-\varepsilon)\pi(\sigma, \sigma^*) + \varepsilon\pi(\sigma, \sigma) \,. \quad (8.5)$$

Converse. If condition 1 holds, then Equation 8.5 can be satisfied for ε sufficiently small. If condition 2 holds, then Equation 8.5 is satisfied for all $0 < \varepsilon < 1$.

Direct. Suppose that $\pi(\sigma^*, \sigma^*) < \pi(\sigma, \sigma^*)$, then $\exists \varepsilon$ sufficiently small that Equation 8.5 is violated. So we have

$$(8.5) \implies \pi(\sigma^*, \sigma^*) \geq \pi(\sigma, \sigma^*) \,.$$

If $\pi(\sigma^*, \sigma^*) = \pi(\sigma, \sigma^*)$, then

$$(8.5) \implies \pi(\sigma^*, \sigma) > \pi(\sigma, \sigma) \,.$$

□

Remark 8.15

The Nash equilibrium condition is $\pi(\sigma^*, \sigma^*) \geq \pi(\sigma, \sigma^*) \ \forall \sigma \neq \sigma^*$ so the condition $\pi(\sigma^*, \sigma) > \pi(\sigma, \sigma)$ in (2) is a supplementary requirement that eliminates some Nash equilibria from consideration. In other words, there may be a Nash equilibrium in the two-player game but no corresponding ESS in the population game. The supplementary condition is particularly relevant in the case of mixed-strategy Nash equilibria.

Theorem 8.14 gives us an alternative procedure for finding an ESS in a pairwise contest population game:

1. write down the associated two-player game;
2. find the symmetric Nash equilibria of this game;
3. test the Nash equilibria using conditions (1) and (2) above.

Any Nash equilibrium strategy σ^* that passes these tests is an ESS, leading to a population profile $\mathbf{x}^* = \sigma^*$.

8.5 ESSs and Nash Equilibria

Example 8.16

Consider The Hawk-Dove game again. The associated two-player game is

		Player 2	
		H	D
Player 1	H	$\frac{v-c}{2}, \frac{v-c}{2}$	$v, 0$
	D	$0, v$	$\frac{v}{2}, \frac{v}{2}$

It is easy to see that (for $v < c$) there are no symmetric pure-strategy Nash equilibria. To find a mixed-strategy Nash equilibrium, we use the Equality of Payoffs theorem (Theorem 4.27)

$$\pi_1(H, \sigma^*) = \pi_1(D, \sigma^*)$$
$$\iff q^* \frac{v-c}{2} + (1-q^*)v = (1-q^*)\frac{v}{2}$$
$$\iff q^* = \frac{v}{c}.$$

By the symmetry of the problem, we can deduce immediately that player 1 also plays H with probability $p^* = \frac{v}{c}$. To show that $\sigma^* = (p^*, 1-p^*)$ is an ESS, we must show that either condition (1) or condition (2) of Theorem 8.14 holds for every $\sigma \neq \sigma^*$. Because σ^* is a mixed strategy, the Equality of Payoffs theorem also tells us that that $\pi(\sigma^*, \sigma^*) = \pi(\sigma, \sigma^*)$. So condition (1) does *not* hold, and the ESS condition becomes

$$\pi(\sigma^*, \sigma) > \pi(\sigma, \sigma).$$

Now

$$\pi(\sigma^*, \sigma) = p^* p \frac{v-c}{2} + p^*(1-p)v + (1-p^*)(1-p)\frac{v}{2}$$

and

$$\pi(\sigma, \sigma) = p^2 \frac{v-c}{2} + p(1-p)v + (1-p)^2 \frac{v}{2}.$$

So, after a few lines of algebra, we find

$$\pi(\sigma^*, \sigma) - \pi(\sigma, \sigma) = \frac{c}{2}(p^* - p)^2$$
$$> 0 \quad \forall p \neq p^*$$

which proves that σ^* is an ESS.

Exercise 8.6

Find the ESSs for the population games defined by the following two-player games.

(a)

	P_2		
P_1	R	G	B
R	1,1	0,0	0,0
G	0,0	1,1	0,0
B	0,0	0,0	1,1

(b)

	P_2	
P_1	G	H
G	3,3	2,2
H	2,2	1,1

(c)

	P_2	
P_1	A	B
A	4,4	0,1
B	1,0	2,2

(d)

	P_2	
P_1	H	D
H	$-\frac{1}{2},-\frac{1}{2}$	2,0
D	0,2	1,1

Exercise 8.7

A population of birds is distributed so that in any given area there are only two females and two trees suitable for nesting (T_1 and T_2). If the two females pick the same nesting site, then they each raise 2 offspring. If they choose different sites, then they are more vulnerable to predators and only raise 1 offspring each. This situation can be modelled as a pairwise contest game. (a) Construct the 2-player payoff table and find all the symmetric Nash equilibria of this game. (b) Determine which of the Nash equilibria correspond to ESSs in the associated population game.

Example 8.17

In Section 6.4, we saw that there is a symmetric, mixed-strategy Nash equilibrium (σ^*, σ^*) for the War of Attrition game. An individual following this strategy will persist for a time drawn at random from an exponential distribution with mean v/c (or, equivalently, will accept a cost $x = ct$ drawn at random from an exponential distribution with mean v). Is the strategy σ^* an ESS?

Because the Nash equilibrium is mixed, we have

$$\pi(\sigma, \sigma^*) = \pi(\sigma^*, \sigma^*) \quad \forall \sigma$$

(after all, that's how the equilibrium was found). So we need to show that

$$\pi(\sigma^*, \sigma) > \pi(\sigma, \sigma) \quad \forall \sigma$$

where σ is *any* pure or mixed strategy. This seems like a tall order, but fortunately the task is easier than it appears. If we can show that

$$\pi(\sigma^*, x) - \pi(x, x) > 0 \quad \forall x \in [0, \infty)$$

then it follows that

$$\pi(\sigma^*, \sigma) - \pi(\sigma, \sigma) = \mathbb{E}_x\left[\pi(\sigma^*, x) - \pi(x, x)\right] > 0 \quad \forall \sigma.$$

Now

$$\begin{aligned}\pi(\sigma^*, x) &= (v - x) \int_x^\infty \frac{1}{v} \exp\left(-\frac{y}{v}\right) dy - \int_0^x y \frac{1}{v} \exp\left(-\frac{y}{v}\right) dy \\ &= 2v \exp\left(-\frac{x}{v}\right) - v\end{aligned}$$

and $\pi(x, x) = -x$. So

$$\pi(\sigma^*, x) - \pi(x, x) = 2v \exp\left(-\frac{x}{v}\right) - v + x$$

which has a minimum value of $v \ln(2)$ at $x = v \ln(2)$. This proves that σ^* is an ESS.

8.6 Asymmetric Pairwise Contests

There are many situations in which the players engaged in a contest can be distinguished. In economic contexts, they may be a buyer and a seller or they may be a firm holding a monopoly in a market and a firm seeking to enter that market. In biological problems, they may be male and female birds dividing up the care of their offspring or they may be larger and smaller stags competing for dominance over a harem of females. Such differences between individuals may lead to an asymmetric payoff table: players may have different actions available to them or the payoffs may differ according to whether the player is male or female, a buyer or a seller. But even if no such payoff asymmetries arise, the possibility that players can occupy different rôles in a game presents us with a problem. For example, it may be reasonable for a male to do one thing and a female to do another. How do we allow for such behaviour given that our formulation of evolutionary stability requires a symmetric game?

In a population, an individual may find themselves playing a particular rôle in one game and playing another rôle in a later encounter. Thus a general strategy must specify behaviour for *all* rôles: use s in rôle r, use s' in rôle r', and so on. By specifying rôle-conditioned strategies, we obtain a symmetric population game. (At first sight, it may seem strange to specify strategies like "care for offspring if male, leave if female" because any given individual is usually either male or female throughout its entire life. In genetic terms, however, the genes that are assumed to control behaviour will be passed down to offspring that may be male or female, whatever the sex of the parent.) Payoffs can be calculated if we know how often an individual assumes a particular rôle and how often the individuals who meet are playing in any two specified rôles. For simplicity, we will assume that there are only two rôles of interest in any game and that a player in one rôle always meets a player in the other rôle. Such a game is said to possess "rôle asymmetry". We will also assume, as is often the case, that each individual finds themselves playing each rôle with equal probability. For example, in a contest between males and females, a gene has a 50% chance of finding itself controlling the behaviour of a male body, if the sex ratio is 1:1.

Example 8.18

Consider a variation on the Hawk-Dove game in which two individuals are contesting ownership of a territory that one of them currently controls. We assume that the value of the territory and the costs of contest are the same for both players. The difference with the standard Hawk-Dove game is that players can now condition their behaviour on the rôle that they occupy – "owner" or "intruder". Therefore, pure strategies are now of the form "play Hawk if owner, play Dove if intruder", which we will represent by the pair of letters HD (a strategy that is often called "Bourgeois"). The full set of pure strategies is HH, HD, DH, and DD.[6] We assume that any contest involves one player in each rôle and that each player has an equal chance of being an owner or an intruder. (In genetic terms, the genes that are currently in an owner may find themselves passed on to an offspring that has yet to find a resource to control.) With these assumptions, we can derive the payoff table shown in Figure 8.1. For example, consider the expected payoff to players using HH against opponents who use HD. Half the time they will be the owner using H against an intruder who uses D, and half the time they will be an intruder using H against an owner who also uses H. The expected payoff is

$$\frac{1}{2}v + \frac{1}{2}\frac{v-c}{2} = \frac{3v-c}{4}.$$

There are two symmetric pure-strategy Nash equilibria: $[HD, HD]$ and $[DH, DH]$. Because (for $v < c$)

$$\frac{v}{2} > \frac{3v-c}{4} > \frac{v}{4} > \frac{2v-c}{4}$$

the strategies HD and DH are both ESSs. There is no mixed strategy ESS (see Exercise 8.8).

---

[6] Surprisingly, the rather bizarre strategy "play Dove if owner, play Hawk if intruder" is found in Nature (though rarely). See Maynard Smith (1982).

8.6 Asymmetric Pairwise Contests

	Player 2			
	HH	HD	DH	DD
HH	$\frac{v-c}{2}, \frac{v-c}{2}$	$\frac{3v-c}{4}, \frac{v-c}{4}$	$\frac{3v-c}{4}, \frac{v-c}{4}$	$v, 0$
HD	$\frac{v-c}{4}, \frac{3v-c}{4}$	$\frac{v}{2}, \frac{v}{2}$	$\frac{2v-c}{4}, \frac{2v-c}{4}$	$\frac{3v}{4}, \frac{v}{4}$
DH	$\frac{v-c}{4}, \frac{3v-c}{4}$	$\frac{2v-c}{4}, \frac{2v-c}{4}$	$\frac{v}{2}, \frac{v}{2}$	$\frac{3v}{4}, \frac{v}{4}$
DD	$0, v$	$\frac{v}{4}, \frac{3v}{4}$	$\frac{v}{4}, \frac{3v}{4}$	$\frac{v}{2}, \frac{v}{2}$

(Player 1 labels the rows.)

Figure 8.1 Payoff table for the asymmetric Hawk-Dove game of Example 8.18.

Exercise 8.8

Set $v = 4$ and $c = 8$ in the payoff table shown in Figure 8.1 and show that there is no mixed strategy ESS.

The absence of mixed strategy ESSs is a general feature of games with rôle asymmetry, as was shown by Selten in 1980. The proof is easier if we consider behavioural rather than mixed strategies. We will consider only games with two rôles and the same two actions in each rôle. In such a game, a general behavioural strategy can be phrased as "use A with probability p_1 in rôle 1, use A with probability p_2 in rôle 2". If we denote a behavioural strategy by β then we can write

$$\beta = (\beta_1, \beta_2)$$

where $\beta_i = (p_i, 1 - p_i)$ is the behaviour specified for rôle i. The payoff for β against β' is then

$$\pi(\beta, \beta') = \frac{1}{2}\pi(\beta_1, \beta'_2) + \frac{1}{2}\pi(\beta_2, \beta'_1).$$

Theorem 8.19

In a pairwise contest game that possesses rôle asymmetry, all evolutionarily stable strategies are pure.

Proof

Suppose that, contrary to the theorem, β^* is a randomising behavioural strategy that is an ESS. Then, by the equality of payoffs theorem, there is another strategy $\hat{\beta}$ for which $\pi(\hat{\beta}, \beta^*) = \pi(\beta^*, \beta^*)$. In fact, there will be many such strategies. Let us pick a $\hat{\beta}$ that differs from β^* in, say, rôle 1 so that $(\hat{\beta}_1, \hat{\beta}_2) = (\hat{\beta}_1, \beta_2^*)$. Then

$$\begin{aligned}\pi(\hat{\beta}, \beta^*) &= \frac{1}{2}\pi(\hat{\beta}_1, \beta_2^*) + \frac{1}{2}\pi(\hat{\beta}_2, \beta_1^*) \\ &= \frac{1}{2}\pi(\hat{\beta}_1, \beta_2^*) + \frac{1}{2}\pi(\beta_2^*, \beta_1^*)\end{aligned}$$

which, together with the condition $\pi(\hat{\beta}, \beta^*) = \pi(\beta^*, \beta^*)$ implies

$$\pi(\hat{\beta}_1, \beta_2^*) = \pi(\beta_1^*, \beta_2^*).$$

Hence

$$\begin{aligned}\pi(\hat{\beta}, \hat{\beta}) &= \frac{1}{2}\pi(\hat{\beta}_1, \hat{\beta}_2) + \frac{1}{2}\pi(\hat{\beta}_2, \hat{\beta}_1) \\ &= \frac{1}{2}\pi(\hat{\beta}_1, \beta_2^*) + \frac{1}{2}\pi(\beta_2^*, \hat{\beta}_1) \\ &= \frac{1}{2}\pi(\beta_1^*, \beta_2^*) + \frac{1}{2}\pi(\beta_2^*, \hat{\beta}_1) \\ &= \frac{1}{2}\pi(\beta_1^*, \hat{\beta}_2) + \frac{1}{2}\pi(\beta_2^*, \hat{\beta}_1) \\ &= \pi(\beta^*, \hat{\beta})\end{aligned}$$

which contradicts our initial assumption. □

Remark 8.20

A more general version of this theorem – for games with more than two actions and more than two rôles – was established by Selten in 1980. However, it is important to note that it only applies to pairwise contest games. It does not hold in general population games that may have a non-linear population-wide component.

8.7 Existence of ESSs

Unfortunately, it is not true that all games have an ESS, as is demonstrated by the following example.

Example 8.21

Consider the two-player (children's) game "Rock-Scissors-Paper". The children simultaneously make the shape of one of the items with their hand: Rock (R)

8.7 Existence of ESSs

beats Scissors (S); Scissors beat Paper (P); Paper beats Rock. If both players choose the same item, then then game is a draw. One payoff table that corresponds to this game is

	R	S	P
R	0,0	1,−1	−1,1
S	−1,1	0,0	1,−1
P	1,−1	−1,1	0,0

This two-player game has a unique Nash equilibrium $[\sigma^*, \sigma^*]$ with $\sigma^* = \left(\frac{1}{3}, \frac{1}{3}, \frac{1}{3}\right)$ but this strategy is not an ESS in the corresponding population game, because (for example)

$$\pi(\sigma^*, R) = 0 = \pi(R, R) .$$

However, one important class of games always has at least one ESS.

Theorem 8.22

All generic, two-action, symmetric pairwise contests have an ESS.

Proof

A symmetric two-player game has the following form.

		P_2	
		A	B
P_1	A	a,a	b,c
	B	c,b	d,d

By applying affine transformations (see Definition 4.34), we can turn this into the equivalent game

		P_2	
		A	B
P_1	A	$a-c, a-c$	0,0
	B	0,0	$d-b, d-b$

It is easy to see that the ESS conditions given in Theorem 8.14 are unaffected by this transformation.

Because we are considering generic games, we have $a \neq c$ and $b \neq d$. There are three possibilities to consider.

1. If $a - c > 0$, then $\pi(A, A) > \pi(B, A)$ and hence $\sigma_A = (1, 0)$ is an ESS, by condition (1) in Theorem 8.14.

2. If $d - b > 0$, then $\pi(B, B) > \pi(A, B)$ and hence $\sigma_B = (0, 1)$ is an ESS, by condition (1) in Theorem 8.14.

3. If $a - c < 0$ and $d - b < 0$, then there is a symmetric mixed strategy[7] Nash equilibrium $[\sigma^*, \sigma^*]$ with $\sigma^* = (p^*, 1 - p^*)$ and
$$p^* = \frac{d-b}{a-c+d-b}.$$
At this equilibrium, $\pi(\sigma^*, \sigma^*) = \pi(\sigma, \sigma^*)$ for any strategy σ, so we have to consider the inequality in condition (2) of Theorem 8.14. Now
$$\pi(\sigma^*, \sigma) = pp^*(a-c) + (1-p)(1-p^*)(d-b)$$
and
$$\pi(\sigma, \sigma) = p^2(a-c) + (1-p)^2(d-b)$$
so
$$\begin{aligned}\pi(\sigma^*, \sigma) - \pi(\sigma, \sigma) &= p(p^* - p)(a-c) + (1-p)(p - p^*)(d-b) \\ &= (p^* - p)\left[p(a-c+d-b) - (d-b)\right] \\ &= -(a-c+d-b)(p^* - p)^2 \\ &> 0\end{aligned}$$

So σ^* is an ESS.

Hence, there is always an ESS in the pairwise contest population game that corresponds to this two-player game. □

Exercise 8.9

Determine whether the population games defined by the following two-player games have an ESS.

(a)

		P_2	
		A	B
P_1	A	1, 1	1, 1
	B	1, 1	1, 1

(b)

		P_2	
		E	F
P_1	E	1, 1	1, 2
	F	2, 1	0, 0

(c)

		P_2		
		A	B	C
	A	0, 0	1, −1	−3, 3
P_1	B	−1, 1	0, 0	2, −2
	C	3, −3	−2, 2	0, 0

---
[7] See Exercise 4.8.

8.7 Existence of ESSs

Apart from the possibility that an ESS may not exist for a given game, it is also the case that the interesting Nash equilibrium strategies in some dynamic games turn out not to be ESSs. For example, in the Iterated Prisoners' Dilemma, the Nash equilibrium strategies that are introduced to ensure cooperative behaviour – "Tit-for-Tat", "Grim", and similar strategies – are not ESSs. For example, we have[8]

$$\pi_i(\sigma_C, \sigma_C) = \pi_i(\sigma_G, \sigma_C) \text{ and } \pi_i(\sigma_C, \sigma_G) = \pi_i(\sigma_G, \sigma_G)$$

which means the ESS conditions in Theorem 8.14 do not hold. This problem arises because the strategic form of the Iterated Prisoners' Dilemma dynamic game is non-generic and many of the Nash equilibria, therefore, occur in continuous sets that provide the same payoff for all points in the set. Because an ESS must have a greater payoff than any other strategy in *all* nearby populations, none of these Nash equilibrium strategies can be an ESS. The failure of many games to have (interesting) ESSs has led to the search for alternative stability concepts that are weaker: these include Neutral Stability, Evolutionarily Stable Sets, and Limit ESSs. However, none of these concepts has gained universal popularity. So, instead, our focus will now shift from the strategies to the evolution of the population structure itself.

---

[8] See Section 7.2 for definitions of these strategies.

9
Replicator Dynamics

9.1 Evolutionary Dynamics

In the previous chapter, we investigated the concept of an evolutionarily stable strategy. Although this concept implicitly assumes the existence of some kind of evolutionary dynamics, it gives an incomplete description. First, an ESS may not exist – in which case the analysis tells us nothing about the evolution of the system described by the game. Second, the definition of an ESS deals only with monomorphic populations in which every individual uses the same strategy. But, if the ESS is a mixed strategy, then all strategies in the support of the ESS obtain the same payoff as the evolutionarily stable strategy itself. So it is pertinent to ask whether a polymorphic population with the same population profile as that generated by the ESS can also be stable. To address these questions, we will look at a specific type of evolutionary dynamics, called replicator dynamics.

We consider a population in which individuals, called "replicators", exist in several different types. Each type of individual uses a pre-programmed strategy (for the game being considered explicitly) and passes this behaviour to its descendants without modification. In the replicator dynamics, it is assumed that individuals are programmed to use only pure strategies from a finite set $\mathbf{S} = \{s_1, s_2, \ldots, s_k\}$. Let n_i be the number of individuals using s_i, then the total population size is

$$N = \sum_{i=1}^{k} n_i$$

and the proportion of individuals using s_i is

$$x_i = \frac{n_i}{N}.$$

The population state can then be described by a vector $\mathbf{x} = (x_1, x_2, \ldots, x_k)$ (together with the overall size of the population N, which will not interest us). Let β and δ be the background *per capita* birth and death rates in the population. That is, β and δ represent the contributions to the rates of appearance and disappearance of individuals in the population which are independent of the game in question. The background *per capita* rate of change of numbers, $\beta - \delta$, is modified by the payoff for using strategy s_i in the population game under consideration. The rate of change of the number of individuals using s_i is[1]

$$\dot{n}_i = (\beta - \delta + \pi(s_i, \mathbf{x}))n_i$$

and the rate of change of the total population size is given by

$$\begin{aligned}
\dot{N} &= \sum_{i=1}^{k} \dot{n}_i \\
&= (\beta - \delta)\sum_{i=1}^{k} n_i + \sum_{i=1}^{k} \pi(s_i, \mathbf{x})n_i \\
&= (\beta - \delta)N + N\sum_{i=1}^{k} x_i \pi(s_i, \mathbf{x}) \\
&= (\beta - \delta + \bar{\pi}(\mathbf{x}))N.
\end{aligned}$$

where we have defined the *average payoff* in the population by

$$\bar{\pi}(\mathbf{x}) = \sum_{i=1}^{k} x_i \pi(s_i, \mathbf{x}).$$

Thus the population grows or declines exponentially. This may not be very realistic, but we can improve the description by letting β and δ depend on N. So long as the fitness increments $\pi(s_i, \mathbf{x})$ depend only on the proportions x_i and not on the actual numbers n_i, the game dynamics will be unchanged.

From a game-theoretic point of view, we are more interested in how the proportions of each type change over time. Now

$$\dot{n}_i = N\dot{x}_i + x_i \dot{N}$$

so

$$\begin{aligned}
N\dot{x}_i &= \dot{n}_i - x_i \dot{N} \\
&= (\beta - \delta + \pi(s_i, \mathbf{x}))x_i N - x_i(\beta - \delta + \bar{\pi}(\mathbf{x}))N.
\end{aligned}$$

---

[1] We use a dot to denote a time derivative so that, for example, $\dot{x} = dx/dt$.

9.1 Evolutionary Dynamics

Cancelling and dividing by N, we have

$$\dot{x}_i = (\pi(s_i, \mathbf{x}) - \bar{\pi}(\mathbf{x}))x_i. \tag{9.1}$$

In other words, the proportion of individuals using strategy s_i increases (decreases) if its payoff is bigger (smaller) than the average payoff in the population.

Exercise 9.1

Clearly, at any time we should have $\sum_{i=1}^{k} x_i = 1$. Show that if this condition is satisfied at time $t = 0$, then it is satisfied for all $t > 0$.

Exercise 9.2

Show that the evolutionary dynamics is unchanged under an affine transformation of the payoffs, provided the time parameter is scaled appropriately. (An affine transformation changes the payoffs by $\pi \to \lambda \pi + \mu$ where μ is a real number and λ is a positive real number.)

Definition 9.1

A *fixed point* of the replicator dynamics is a population that satisfies $\dot{x}_i = 0$ $\forall i$. Fixed points describe populations that are no longer evolving.

Example 9.2

Consider a pairwise contest population game with action set $A = \{E, F\}$ and payoffs

$$\pi(E,E) = 1 \quad \pi(E,F) = 1 \quad \pi(F,E) = 2 \quad \pi(F,F) = 0.$$

So $\pi(E, \mathbf{x}) = x_1 + x_2$ and $\pi(F, \mathbf{x}) = 2x_1$, which gives

$$\begin{aligned}\bar{\pi}(\mathbf{x}) &= x_1(x_1 + x_2) + x_2(2x_1) \\ &= x_1^2 + 3x_1 x_2.\end{aligned}$$

The replicator dynamics for this game is

$$\begin{aligned}\dot{x}_1 &= x_1(x_1 + x_2 - x_1^2 - 3x_1 x_2) \\ \dot{x}_2 &= x_2(2x_1 - x_1^2 - 3x_1 x_2).\end{aligned}$$

So the fixed points are $(x_1 = 0, x_2 = 1)$, $(x_1 = 1, x_2 = 1)$ and $(x_1 = \frac{1}{2}, x_2 = \frac{1}{2})$.

Exercise 9.3

Consider the pairwise contest with payoffs given in the table below (where $a < b$).

	A	B
A	$a-b, a-b$	$2a, 0$
B	$0, 2a$	a, a

Derive the replicator dynamics equations for this game and find all the fixed points.

9.2 Two-strategy Pairwise Contests

Dealing with general games requires some mathematical techniques that not everyone will be familiar with. So, we will temporarily make a further simplification and consider pairwise contest games that only have two pure strategies. Suppose $\mathbf{S} = \{s_1, s_2\}$ and let $x \equiv x_1$. Then $x_2 = 1 - x$ and $\dot{x}_2 = -\dot{x}_1$. So we only need to consider a single differential equation

$$\dot{x} = (\pi(s_1, \mathbf{x}) - \bar{\pi}(\mathbf{x}))x.$$

We can simplify this further by substituting

$$\bar{\pi}(\mathbf{x}) = x\pi(s_1, \mathbf{x}) + (1-x)\pi(s_2, \mathbf{x})$$

which gives

$$\dot{x} = x(1-x)(\pi(s_1, \mathbf{x}) - \pi(s_2, \mathbf{x})).$$

Example 9.3

Consider a pairwise contest Prisoners' Dilemma. The pure strategies are $\{C, D\}$ and the payoffs to the focal individual in the corresponding 2-player game are $\pi(C, C) = 3$, $\pi(C, D) = 0$, $\pi(D, C) = 5$, and $\pi(D, D) = 1$. Let x be the proportion of individuals using C, then

$$\pi(C, \mathbf{x}) = 3x + 0(1-x) = 3x$$

and

$$\pi(D, \mathbf{x}) = 5x + 1(1-x) = 1 + 4x.$$

The rate of change of the proportion of individuals using C is

$$\begin{aligned}\dot{x} &= x(1-x)(\pi(C, \mathbf{x}) - \pi(D, \mathbf{x})) \\ &= x(1-x)(3x - (1+4x)) \\ &= -x(1-x)(1+x).\end{aligned}$$

9.2 Two-strategy Pairwise Contests

The fixed points for this dynamical system are $x^* = 0$ and $x^* = 1$. We know that the unique Nash equilibrium for the Prisoners' Dilemma game is for everyone to defect (play D). This means that $x^* = 0$ corresponds to a Nash equilibrium but $x^* = 1$ does not. We also see that $\dot{x} < 0$ for $x \in (0, 1)$. This means that any population that is not at a fixed point of the dynamics will evolve *towards* the fixed point that corresponds to the Nash equilibrium and *away* from the other one.

Exercise 9.4

Derive the replicator dynamics for the Hawk-Dove game and show that any population that is not at a fixed point will evolve towards the point that corresponds to the unique symmetric Nash equilibrium.

It seems that every Nash equilibrium corresponds to a fixed point in the replicator dynamics but not every fixed point corresponds to a Nash equilibrium. The following theorem proves this conjecture for pairwise contest games with two pure strategies.

Theorem 9.4

Let $\mathbf{S} = \{s_1, s_2\}$ and let $\sigma^* = (p^*, 1 - p^*)$ be the strategy that uses s_1 with probability p^*. If (σ^*, σ^*) is a symmetric Nash equilibrium, then the population $\mathbf{x}^* = (x^*, 1 - x^*)$ with $x^* = p^*$ is a fixed point of the replicator dynamics $\dot{x} = x(1-x)(\pi(s_1, \mathbf{x}) - \pi(s_2, \mathbf{x}))$.

Proof

If σ^* is a pure strategy, then $x^* = 0$ or $x^* = 1$. In either case, we have $\dot{x} = 0$. If σ^* is a mixed strategy, then Theorem 4.27 says that $\pi(s_1, \sigma^*) = \pi(s_2, \sigma^*)$. Now, for a pairwise contest,

$$\begin{aligned}\pi(s_i, \sigma^*) &= p^* \pi(s_i, s_1) + (1 - p^*) \pi(s_i, s_2) \\ &= \pi(s_i, \mathbf{x}^*) \,.\end{aligned}$$

So we have $\pi(s_1, \mathbf{x}^*) = \pi(s_2, \mathbf{x}^*)$ and consequently $\dot{x} = 0$. □

We have shown that Nash equilibria in two-player games and fixed points in the replicator dynamics are related. Is there a consistent relation between the ESSs in a population game and the fixed points?

Example 9.5

Consider a pairwise contest with actions A and B and the following payoffs in the associated two-player game: $\pi(A, A) = 3$, $\pi(B, B) = 1$ and $\pi(A, B) = \pi(B, A) = 0$. The ESSs are for everyone to play A or for everyone to play B. The mixed strategy $\sigma = (\frac{1}{4}, \frac{3}{4})$ is not an ESS. Let x be the proportion of individuals using A, then the rate of change of the proportion of individuals using A is

$$\begin{aligned}\dot{x} &= x(1-x)(\pi(A, \mathbf{x}) - \pi(B, \mathbf{x})) \\ &= x(1-x)(3x - (1-x)) \\ &= x(1-x)(4x - 1).\end{aligned}$$

The fixed points for this dynamical system are $x^* = 0$, $x^* = 1$ and $x^* = \frac{1}{4}$. However, we can see that $\dot{x} > 0$ if $x > \frac{1}{4}$ and $\dot{x} < 0$ if $x < \frac{1}{4}$, so only the pure-strategy behaviours are evolutionary end points. If the population starts in a state where more than 25% of individuals use strategy A, then the population evolves until everyone uses A. On the other hand, If the population starts in a state where fewer than 25% of individuals use strategy A, then the population evolves until everyone uses B. This means that only the evolutionary end points correspond to an ESS.

In the Hawk-Dove game, the correspondence between the evolutionary end-point of the replicator dynamics and the ESS is a bit less direct (see Exercise 9.4). The ESS is for *each individual* to play Hawk with probability v/c. However, in the replicator dynamics, individuals cannot use mixed strategies: an individual must either be a pre-programmed Hawk-user or a pre-programmed Dove-user. Nevertheless, the population evolves towards a state in which the proportion of Hawk-users is v/c, which is the polymorphic equivalent of the monomorphic ESS.

Exercise 9.5

A population of birds is distributed so that in any given area there are only two females and two trees suitable for nesting (T_1 and T_2). If the two females pick the same nesting site, then they each raise 2 offspring. If they choose different sites, then they are more vulnerable to predators and only raise 1 offspring each. This situation can be modelled as a pairwise contest game. Derive the replicator dynamics equation and show that only the fixed points that correspond to an ESS are evolutionary end points.

9.3 Linearisation and Asymptotic Stability

In the examples considered in the previous section, an ESS always corresponds to an evolutionary end point in the replicator dynamics. Do all ESSs have a corresponding end point and do all evolutionary end points have a corresponding ESS? In this section, we continue to consider the special case of two-strategy pairwise contest games. Later we will consider general n-strategy games and reach a similar conclusion. Because the definition of an ESS considers small deviations from a specified population, it makes sense to do the same in the replicator dynamics.

Definition 9.6

A fixed point of the replicator dynamics (or any dynamical system) is said to be *asymptotically stable* if any small deviations from that state are eliminated by the dynamics as $t \to \infty$.

Example 9.7

Consider a pairwise contest with pure strategies A and B and the following payoffs in the associated two-player game

$$\pi(A,A) = 3 \quad \pi(B,B) = 1 \quad \pi(A,B) = \pi(B,A) = 0.$$

We know that the ESSs for this game are for everyone to play A or for everyone to play B. The mixed strategy $\sigma = (\frac{1}{4}, \frac{3}{4})$ is a Nash equilibrium but it is not an ESS. Let x be the proportion of individuals using A, then the replicator dynamics is

$$\dot{x} = -x(1-x)(1-4x)$$

with fixed points at $x^* = 0$, $x^* = 1$ and $x^* = \frac{1}{2}$.

First, consider a population near to $x^* = 0$. Let $x = x^* + \varepsilon = \varepsilon$ where we must have $\varepsilon > 0$ to ensure $x > 0$. Then $\dot{x} = \dot{\varepsilon}$ because x^* is a constant. Thus we have

$$\dot{\varepsilon} = -\varepsilon(1-\varepsilon)(1-4\varepsilon) \, .$$

Because it is assumed that $\varepsilon \ll 1$, we can ignore terms proportional to ε^n where $n > 1$. This procedure is called *linearisation*. Thus

$$\dot{\varepsilon} \approx -\varepsilon$$

which has the solution

$$\varepsilon(t) = \varepsilon_0 e^{-t}.$$

This tells us that the dynamics reduces small deviations from the population state $\mathbf{x} = (0, 1)$ (i.e., $\varepsilon \to 0$ as $t \to \infty$). In other words, the fixed point $x^* = 0$ is asymptotically stable.

Now consider a population near to $x^* = 1$. Let $x = x^* - \varepsilon = 1 - \varepsilon$ with $\varepsilon > 0$ (to ensure $x < 1$). Following the linearisation procedure we find that

$$\dot{\varepsilon} \approx -3\varepsilon$$

which has the solution

$$\varepsilon(t) = \varepsilon_0 e^{-3t}.$$

i.e., $x^* = 1$ is asymptotically stable.

Finally, consider a population near to $x^* = \frac{1}{4}$. Let $x = x^* + \varepsilon = \frac{1}{4} + \varepsilon$ (with no sign restriction on ε). Then we have

$$\dot{\varepsilon} \approx \frac{1}{16}\varepsilon$$

with solution

$$\varepsilon(t) = \varepsilon_0 e^{t/16}$$

So $x_3^* = \frac{1}{4}$ is *not* asymptotically stable. (In fact, it is unstable.)

So in this case we find that a strategy is an ESS if and only if the corresponding fixed point in the replicator dynamics is asymptotically stable.

Theorem 9.8

For any two-strategy pairwise contest, a strategy is an ESS if and only if the corresponding fixed point in the replicator dynamics is asymptotically stable.

Proof

Consider a pairwise contest with strategies A and B. Let x be the proportion of individuals using A, then the replicator dynamics is given by

$$\dot{x} = x(1-x)[\pi(A, \mathbf{x}) - \pi(B, \mathbf{x})].$$

There are three possible cases to consider: a single pure-strategy ESS or stable monomorphic population; two pure-strategy ESSs or stable monomorphic populations; and one mixed strategy ESS or polymorphic population.

1. Let $\sigma^* = (1, 0)$. Then (for $\sigma = (y, 1-y)$ with $y \neq 1$) σ^* is an ESS if and only if

$$\pi(A, \mathbf{x}_\varepsilon) - \pi(\sigma, \mathbf{x}_\varepsilon) > 0$$
$$\iff \pi(A, \mathbf{x}_\varepsilon) - y\pi(A, \mathbf{x}_\varepsilon) - (1-y)\pi(B, \mathbf{x}_\varepsilon) > 0$$
$$\iff (1-y)[\pi(A, \mathbf{x}_\varepsilon) - \pi(B, \mathbf{x}_\varepsilon)] > 0$$
$$\iff \pi(A, \mathbf{x}_\varepsilon) - \pi(B, \mathbf{x}_\varepsilon) > 0$$

9.3 Linearisation and Asymptotic Stability

Let $x = 1 - \varepsilon$ with $\varepsilon > 0$. Then

$$\dot{\varepsilon} = -\varepsilon[\pi(A, \mathbf{x}_\varepsilon) - \pi(B, \mathbf{x}_\varepsilon)].$$

So $\sigma^* = (1, 0)$ is an ESS if and only if the corresponding population $x^* = 1$ is asymptotically stable.

2. Let $\sigma^* = (0, 1)$. Then, using a similar argument to the previous case, σ^* is an ESS if and only if

$$\pi(A, \mathbf{x}_\varepsilon) - \pi(B, \mathbf{x}_\varepsilon) < 0.$$

Let $x = \varepsilon$ with $\varepsilon > 0$. Then

$$\dot{\varepsilon} = \varepsilon[\pi(A, \mathbf{x}_\varepsilon) - \pi(B, \mathbf{x}_\varepsilon)].$$

So $\sigma^* = (0, 1)$ is an ESS if and only if the corresponding population $x^* = 0$ is asymptotically stable.

3. Let $\sigma^* = (p^*, 1 - p^*)$ with $0 < p^* < 1$. Then σ^* is an ESS if and only if $\pi(\sigma^*, \sigma) > \pi(\sigma, \sigma)$. Taking $\sigma = A$ and $\sigma = B$ in turn, this condition becomes the two conditions

$$\pi(A, A) < \pi(B, A) \quad \text{and} \quad \pi(B, B) < \pi(A, B).$$

Let $x = x^* + \varepsilon$. Then, for a pairwise contest, the replicator dynamics equation

$$\dot{x} = x(1 - x)[\pi(A, \mathbf{x}_\varepsilon) - \pi(B, \mathbf{x}_\varepsilon)]$$

becomes

$$\dot{\varepsilon} = x^*(1 - x^*)\varepsilon\left([\pi(A, A) - \pi(B, A)] + [\pi(B, B) - \pi(B, A)]\right)$$

using the assumption that $\mathbf{x}^*$ is a fixed point. So $\mathbf{x}^*$ is asymptotically stable if and only if σ^* is an ESS.

□

Let $\mathbf{F}$ be the set of fixed points and let $\mathbf{A}$ be the set of asymptotically stable fixed points in the replicator dynamics. Let $\mathbf{N}$ be the set of symmetric Nash equilibrium strategies and let $\mathbf{E}$ be the set of ESSs in the symmetric game corresponding to the replicator dynamics. Then we have shown that, for any *two-strategy pairwise-contest game*, the following relationships hold for a strategy σ^* and the corresponding population state $\mathbf{x}^*$:

1. $\sigma^* \in \mathbf{E} \iff \mathbf{x}^* \in \mathbf{A}$;

2. $\mathbf{x}^* \in \mathbf{A} \implies \sigma^* \in \mathbf{N}$;[2]

3. $\sigma^* \in \mathbf{N} \implies \mathbf{x}^* \in \mathbf{F}$.

Allowing our now customary abuse of notation that identifies a strategy with its corresponding population state, we can write these relations more concisely as

$$\mathbf{E} = \mathbf{A} \subseteq \mathbf{N} \subseteq \mathbf{F}.$$

As we shall see, for pairwise-contest games with more than two strategies these relations become

$$\mathbf{E} \subseteq \mathbf{A} \subseteq \mathbf{N} \subseteq \mathbf{F}.$$

Exercise 9.6

Consider the pairwise contest with payoffs given in the table below (where $a \neq 0$).

	A	B
A	a,a	0,0
B	0,0	a,a

Find all the ESSs of this game for the cases $a > 0$ and $a < 0$. Derive the replicator dynamics equation for the proportion of A-players x. Find all the fixed points of the replicator dynamics equation (for $a > 0$ and $a < 0$). Show that only the fixed points that correspond to an ESS are asymptotically stable.

9.4 Games with More Than Two Strategies

If we increase the number of pure strategies to n, then we have n equations to deal with.

$$\dot{x}_i = f_i(\mathbf{x}) \quad i = 1, \ldots, n.$$

Using the constraint $\sum_{i=1}^{n} x_i = 1$, we can introduce a reduced state vector $(x_1, x_2, \ldots, x_{n-1})$ and reduce the number of equations to $n - 1$.

$$\dot{x}_i = f_i(x_1, x_2, \ldots, x_{n-1}) \quad i = 1, \ldots, n - 1.$$

---
[2] This follows from the first equivalence because $\sigma^* \in \mathbf{E} \implies \sigma^* \in \mathbf{N}$.

9.4 Games with More Than Two Strategies

We can write this dynamical system more compactly in vector format as[3]

$$\dot{\mathbf{x}} = \mathbf{f}(\mathbf{x}) \ .$$

There is no confusion introduced by referring to both types of state with the same symbol, $\mathbf{x}$. It will be clear from the context which type of state is being referred to.

Example 9.9

Consider the following pairwise contest game, which will be used as the basis of all the examples in this section. The game has the payoff table

	A	B	C
A	0,0	3,3	1,1
B	3,3	0,0	1,1
C	1,1	1,1	1,1

The replicator dynamics for this game is

$$\begin{aligned}
\dot{x}_1 &= x_1(3x_2 + x_3 - \bar{\pi}(\mathbf{x})) \\
\dot{x}_2 &= x_2(3x_1 + x_3 - \bar{\pi}(\mathbf{x})) \\
\dot{x}_3 &= x_3(1 - \bar{\pi}(\mathbf{x}))
\end{aligned}$$

with $\bar{\pi}(\mathbf{x}) = 6x_1x_2 + x_1x_3 + x_2x_3 + x_3$. Writing $x_1 = x$, $x_2 = y$ and $x_3 = 1-x-y$, this system can be reduced to the two-variable dynamical system

$$\begin{aligned}
\dot{x} &= x(1 - x + 2y - \bar{\pi}(x,y)) \\
\dot{y} &= y(1 + 2x - y - \bar{\pi}(x,y))
\end{aligned}$$

with $\bar{\pi}(x,y) = 1 + 4xy - x^2 - y^2$.

Exercise 9.7

Find all the Nash equilibria and ESSs for the game in Example 9.9. Show that the set of fixed points for the replicator dynamics is the same whether we consider the full or the reduced system.

---

[3] Readers who are not familiar with basic dynamical systems theory may find it useful to read Appendix B.

Definition 9.10

The replicator dynamics is defined on the simplex

$$\Delta = \left\{ x_1, x_2, \ldots, x_n \mid 0 \leq x_i \leq 1 \; \forall i \; \& \; \sum_{i=1}^{n} x_i = 1 \right\}.$$

An *invariant manifold* is a connected subset $M \subset \Delta$ such that if $\mathbf{x}(0) \in M$, then $\mathbf{x}(t) \in M$ for all $t > 0$.

It follows immediately from the definition that the fixed points of a dynamical system are invariant manifolds. Boundaries of the simplex Δ (subsets where one or more population types are absent) are also invariant because $x_i = 0 \implies \dot{x}_i = 0$.

Example 9.11

For the dynamical system

$$\begin{aligned} \dot{x} &= x(1 - x + 2y - \bar{\pi}(x,y)) \\ \dot{y} &= y(1 + 2x - y - \bar{\pi}(x,y)) \end{aligned}$$

the obvious invariant manifolds are the fixed points (see previous exercise) and the boundary lines $x = 0$ and $y = 0$. The boundary line $x + y - 1 = 0$ is invariant because (on that line)

$$\begin{aligned} \frac{d}{dt}(x+y) &= \dot{x} + \dot{y} \\ &= (x + y - 1)(1 - \bar{\pi}(x,y)) \\ &= 0 \end{aligned}$$

The line $x = y$ is also invariant because $\dot{x} = \dot{y}$ on that line.

To obtain a qualitative picture of the solutions of the dynamical system, we consider the behaviour of the solutions on (or close to) the invariant manifolds. First, let us consider a fixed point $\mathbf{x}^*$. By making a Taylor expansion of the dynamical system about this fixed point, we obtain a linear approximation to the dynamical system (remember that $\mathbf{f}(\mathbf{x}^*) = 0$ so the constant term vanishes):

$$\dot{x}_i = \sum_{j=1}^{n-1} (x_j - x_j^*) \frac{\partial f_i}{\partial x_j}(\mathbf{x}^*).$$

9.4 Games with More Than Two Strategies

Defining $\xi_i = x_i - x_i^*$, we have

$$\dot{\xi}_i = \sum_{j=1}^{n-1} \xi_j \frac{\partial f_i}{\partial x_j}(\mathbf{x}^*)$$

which is a linear system $\dot{\xi} = L\xi$ with a fixed point at the origin. The matrix L has constant components

$$L_{ij} = \frac{\partial f_i}{\partial x_j}(\mathbf{x}^*)$$

and its eigenvalues determine the behaviour of the linearised system at the fixed point. Provided the fixed point is hyperbolic (i.e., all eigenvalues have non-zero real part), the behaviour of the full, non-linear system is the same.[4] Combining this information with the behaviour of solutions on the other invariant manifolds is usually sufficient to determine a complete qualitative picture of the solutions to the dynamical system.

Example 9.12

Returning to our example, let us consider the fixed point $(x^*, y^*) = (\frac{1}{2}, \frac{1}{2})$. Close to this point we have the linear approximation

$$\begin{pmatrix} \dot{\xi} \\ \dot{\eta} \end{pmatrix} = \begin{pmatrix} -1 & \frac{1}{2} \\ \frac{1}{2} & -1 \end{pmatrix} \begin{pmatrix} \xi \\ \eta \end{pmatrix}$$

The eigenvalues are found from the characteristic equation $\det(L - \lambda I) = 0$, which yields $\lambda_1 = -\frac{1}{2}$ and $\lambda_2 = -\frac{3}{2}$. Because the real parts of both eigenvalues are negative, the fixed point is a stable node. Solving the eigenvector equation

$$\begin{pmatrix} -1 & \frac{1}{2} \\ \frac{1}{2} & -1 \end{pmatrix} \begin{pmatrix} \xi \\ \eta \end{pmatrix} = \lambda \begin{pmatrix} \xi \\ \eta \end{pmatrix}$$

gives the eigenvectors corresponding to each eigenvalue. In this case, we find the eigenvector corresponding to $\lambda = -\frac{3}{2}$ is $\xi = -\eta$, which lies along the boundary line $x + y = 1$. The eigenvector corresponding to $\lambda = -\frac{1}{2}$ is $\xi = \eta$, which lies along the line $x = y$. This eigenvector also passes through the fixed point $(x^*, y^*) = (0,0)$, which is a good indication that the line $x = y$ might be invariant for this dynamical system – as, indeed, we have already shown that it is.

The fixed points $(x^*, y^*) = (1, 0)$ and $(x^*, y^*) = (0, 1)$ both have eigenvalues $\lambda_1 = 3$ and $\lambda_2 = 1$, so both points are unstable nodes. Close to the point $(x^*, y^*) = (0, 0)$ the linear approximation is

$$\begin{pmatrix} \dot{\xi} \\ \dot{\eta} \end{pmatrix} = \begin{pmatrix} 0 & 0 \\ 0 & 0 \end{pmatrix} \begin{pmatrix} \xi \\ \eta \end{pmatrix}$$

---

[4] This is the Hartman-Grobman theorem. See Appendix B.

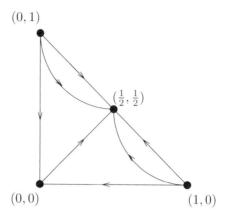

Figure 9.1 The behaviour of the replicator dynamics system from Example 9.9 can be constructed by linking the fixed points by smooth trajectories that are consistent with behaviour of the system on or near to the invariant manifolds (fixed points and invariant lines).

which is not hyperbolic ($\lambda_1 = \lambda_2 = 0$). So the linearisation tells us nothing about the stability properties of this fixed point.

Let us now look at the behaviour of the system on the invariant lines. On the line $y = 0$ we have $\dot{x} = x^2(x-1)$, so $\dot{x} < 0$ for $0 < x < 1$. Similarly, on the line $x = 0$ we have $\dot{y} < 0$ for $0 < y < 1$. On the line $x = y$ we have $\dot{x} = x^2(1-2x)$, so x and y are both increasing for $0 < x, y < \frac{1}{2}$. On the line $x + y - 1 = 0$ we have

$$\begin{aligned}\dot{x} &= x(3 - 3x - \bar{\pi}(x, 1-x)) \\ &= x(3 - 9x + 6x^2).\end{aligned}$$

Hence x is increasing (y is decreasing) for $0 < x < \frac{1}{2}$ and x is decreasing (y is increasing) for $\frac{1}{2} < x < 1$.[5]

Combining all this information we can produce the qualitative picture of the dynamics shown in Figure 9.1.

Exercise 9.8

Draw a qualitative picture of the replicator dynamics for the pairwise contest game with payoff table shown below.

---

[5] Because the point $\mathbf{x}^* = (\frac{1}{2}, \frac{1}{2})$ is a stable node, any other behaviour would indicate that a mistake had been made somewhere.

9.5 Equilibria and Stability

	A	B	C
A	3,3	0,0	1,1
B	0,0	3,3	1,1
C	1,1	1,1	1,1

9.5 Equilibria and Stability

Let **F** be the set of fixed points and let **A** be the set of asymptotically stable fixed points in the replicator dynamics. Let **N** be the set of (symmetric) Nash equilibrium strategies and let **E** be the set of ESSs in the symmetric game corresponding to the replicator dynamics. We will show that, for any *pairwise-contest game*, the following relationships hold for a strategy σ^* and the corresponding population state $\mathbf{x}^*$:

1. $\sigma^* \in \mathbf{E} \Longrightarrow \mathbf{x}^* \in \mathbf{A}$;
2. $\mathbf{x}^* \in \mathbf{A} \Longrightarrow \sigma^* \in \mathbf{N}$;
3. $\sigma^* \in \mathbf{N} \Longrightarrow \mathbf{x}^* \in \mathbf{F}$.

Allowing our customary abuse of notation that identifies a strategy with its corresponding population state, we can write these relations more concisely as

$$\mathbf{E} \subseteq \mathbf{A} \subseteq \mathbf{N} \subseteq \mathbf{F}.$$

First we consider the inclusion $\mathbf{N} \subseteq \mathbf{F}$.

Theorem 9.13

If (σ^*, σ^*) is a symmetric Nash equilibrium, then the population state $\mathbf{x}^* = \sigma^*$ is a fixed point of the replicator dynamics.

Proof

Suppose the Nash equilibrium strategy σ^* is pure, so that every player in the population uses some strategy s_j. Then $x_i = 0$ for $i \neq j$ and $\bar{\pi}(\mathbf{x}^*) = \pi(s_j, \mathbf{x}^*)$. Hence $\dot{x}_i = 0 \ \forall i$.

Suppose the Nash equilibrium strategy σ^* is mixed and let $\mathbf{S}^*$ be the support of σ^* (i.e., $\mathbf{S}^*$ contains only those pure strategies that are played with non-zero probability under σ^*). The equality of payoffs theorem (Theorem 4.27) gives

$$\pi(s, \sigma^*) = \pi(\sigma^*, \sigma^*) \quad \forall s \in \mathbf{S}^*$$

This implies that, in a polymorphic population with $\mathbf{x}^* = \sigma^*$, we must have for all $s_i \in \mathbf{S}^*$

$$\begin{aligned}\pi(s_i, \mathbf{x}^*) &= \sum_{j=1}^{k} \pi(s_i, s_j) x_j \\ &= \sum_{j=1}^{k} \pi(s_i, s_j) p_j \\ &= \pi(s_i, \sigma^*) \\ &= \text{constant}\end{aligned}$$

For strategies $s_i \notin \mathbf{S}^*$, the condition $\mathbf{x}^* = \sigma^*$ gives us $x_i = 0$ and hence $\dot{x}_i = 0$. For strategies $s_j \in \mathbf{S}^*$ we have

$$\begin{aligned}\dot{x}_j &= x_j \left(\pi(s_j, \mathbf{x}^*) - \sum_{i=1}^{k} x_i \pi(s_j, \mathbf{x}^*) \right) \\ &= x_j \left(\pi(s_j, \mathbf{x}^*) - \pi(s_j, \mathbf{x}^*) \sum_{j=1}^{k} x_j \right) \\ &= 0.\end{aligned}$$

□

Remark 9.14

Theorem 9.13 shows that an evolutionary process can produce apparently rational (Nash equilibrium) behaviour in a population composed of individuals who are not required to make consciously rational decisions. In populations where the agents are assumed to have some critical faculties – such as human populations – the requirements of rationality are much less stringent than they are in classical game theory. Individuals are no longer required to be able to work through the (possibly infinite) sequence of reaction and counter-reaction to changes in behaviour. They merely have to be able to evaluate the consequences of their actions, compare them to the results obtained by others who behaved differently and swap to a better (not necessarily the *best*) strategy for the current situation. The population is stable when, given what everyone else is doing, no individual would get a better result by adopting a different strategy. This population view of a Nash equilibrium was first advanced by Nash himself, who called it the "mass action" interpretation.

Next we consider the inclusion $\mathbf{A} \subseteq \mathbf{N}$.

Theorem 9.15

If $\mathbf{x}^*$ is an asymptotically stable fixed point of the replicator dynamics, then the symmetric strategy pair $[\sigma^*, \sigma^*]$ with $\sigma^* = \mathbf{x}^*$ is a Nash equilibrium.

Proof

First, we observe that if $\mathbf{x}^*$ is a fixed point with $x_i > 0\ \forall i$ (i.e., all pure strategy types are present in the population), then all pure strategies must earn the same payoff in that population. It follows from the correspondence of σ^* and $\mathbf{x}^*$ that $\pi_1(s, \sigma^*) = \pi(s, \mathbf{x}^*)$ is also constant for all pure strategies s. Therefore, $[\sigma^*, \sigma^*]$ is a Nash equilibrium.

It remains for us to consider stationary populations where one or more pure strategy types are absent. Denote the set of pure strategies that are present by $S^* \subset S$ (i.e., S^* is the support of the fixed point $\mathbf{x}^*$ and the postulated Nash equilibrium strategy σ^*). Because $\mathbf{x}^*$ is a fixed point, we must have $\pi(s, \mathbf{x}^*) = \bar{\pi}(\mathbf{x}^*)\ \forall s \in S^*$ and $\pi_1(s, \sigma^*) = \pi_1(\sigma^*, \sigma^*)\ \forall s \in S^*$. Now suppose that $[\sigma^*, \sigma^*]$ is not a Nash equilibrium. Then there must be some strategy $s' \notin S^*$ for which $\pi_1(s', \sigma^*) > \pi_1(\sigma^*, \sigma^*)$ and consequently for which $\pi(s', \mathbf{x}^*) > \bar{\pi}(\mathbf{x}^*)$. Consider a population $\mathbf{x}_\varepsilon$ that is close to the state $\mathbf{x}^*$ but has a small proportion ε of s' players. Then

$$\begin{aligned}\dot{\varepsilon} &= \varepsilon\left(\pi(s', \mathbf{x}_\varepsilon) - \bar{\pi}(\mathbf{x}_\varepsilon)\right) \\ &= \varepsilon\left(\pi(s', \mathbf{x}^*) - \bar{\pi}(\mathbf{x}^*)\right) + O(\varepsilon^2).\end{aligned}$$

So the proportion of s'-players increases, contradicting the assumption that $\mathbf{x}^*$ is asymptotically stable. $\square$

Finally we consider the inclusion $\mathbf{E} \subseteq \mathbf{A}$.

Definition 9.16

Let $\dot{\mathbf{x}} = \mathbf{f}(\mathbf{x})$ be a dynamical system with a fixed point at $\mathbf{x}^*$. Then a scalar function $V(\mathbf{x})$, defined for allowable states of the system close to $\mathbf{x}^*$, such that

1. $V(\mathbf{x}^*) = 0$
2. $V(\mathbf{x}) > 0$ for $\mathbf{x} \neq \mathbf{x}^*$
3. $\frac{dV}{dt} < 0$ for $\mathbf{x} \neq \mathbf{x}^*$

is called a (strict) *Lyapounov function*. If such a function exists, then the fixed point $\mathbf{x}^*$ is asymptotically stable.

Theorem 9.17

Every ESS corresponds to an asymptotically stable fixed point in the replicator dynamics. That is, if σ^* is an ESS, then the population with $\mathbf{x}^* = \sigma^*$ is asymptotically stable.

Proof

If σ^* is an ESS then, by definition, there exists an $\bar{\varepsilon}$ such that for all $\varepsilon < \bar{\varepsilon}$

$$\pi(\sigma^*, \sigma_\varepsilon) > \pi(\sigma, \sigma_\varepsilon) \quad \forall \sigma \neq \sigma^*$$

where $\sigma_\varepsilon = (1-\varepsilon)\sigma^* + \varepsilon\sigma'$. In particular, this holds for $\sigma = \sigma_\varepsilon$, so $\pi(\sigma^*, \sigma_\varepsilon) > \pi(\sigma_\varepsilon, \sigma_\varepsilon)$. This implies that in the replicator dynamics we have, for $x^* = \sigma^*$, $x = (1-\varepsilon)x^* + \varepsilon x'$ and all $\varepsilon < \bar{\varepsilon}$

$$\pi(\sigma^*, x) > \bar{\pi}(x).$$

Now consider the relative entropy function

$$V(\mathbf{x}) = -\sum_{i=1}^{k} x_i^* \ln\left(\frac{x_i}{x_i^*}\right)$$

Clearly $V(\mathbf{x}^*) = 0$ and (using Jensen's inequality $\mathbb{E} f(x) \geq f(\mathbb{E} x)$ for any convex function, such as a logarithm)

$$\begin{aligned} V(\mathbf{x}) &= -\sum_{i=1}^{k} x_i^* \ln\left(\frac{x_i}{x_i^*}\right) \\ &\geq -\ln\left(\sum_{i=1}^{k} x_i^* \frac{x_i}{x_i^*}\right) \\ &= -\ln\left(\sum_{i=1}^{k} x_i\right) \\ &= -\ln(1) \\ &= 0. \end{aligned}$$

The time derivative of $V(\mathbf{x})$ along solution trajectories of the replicator dynamics is

$$\begin{aligned} \frac{d}{dt}V(\mathbf{x}) &= \sum_{i=1}^{k} \frac{\partial V}{\partial x_i} \dot{x}_i \\ &= -\sum_{i=1}^{k} \frac{x_i^*}{x_i} \dot{x}_i \end{aligned}$$

9.5 Equilibria and Stability

$$= -\sum_{i=1}^{k} \frac{x_i^*}{x_i} x_i(\pi(s_i, x) - \bar{\pi}(x))$$
$$= -[\pi(\sigma^*, x) - \bar{\pi}(x)].$$

If σ^* is an ESS, then we established above that there is a region near to $\mathbf{x}^*$ where $[\pi(\sigma^*, x) - \bar{\pi}(x)] > 0$ for $\mathbf{x} \neq \mathbf{x}^*$. Hence

$$\frac{dV}{dt} < 0$$

for population states sufficiently close to the fixed point. $V(\mathbf{x})$ is therefore a strict Lyapounov function in this region, and the fixed point $\mathbf{x}^*$ is asymptotically stable. □

The three preceding theorems establish the advertised relationship between the sets of ESSs (**E**), symmetric Nash equilibria (**N**), fixed points (**F**), and asymptotically stable fixed points (**A**):

$$\mathbf{E} \subseteq \mathbf{A} \subseteq \mathbf{N} \subseteq \mathbf{F}.$$

In general, there may be asymptotically stable fixed points in the replicator dynamics which do not correspond to an ESS as is shown in the next exercise.

Exercise 9.9

Consider the pairwise contest game with the payoff table below. Show that the polymorphic population $\mathbf{x}^* = \left(\frac{1}{3}, \frac{1}{3}, \frac{1}{3}\right)$ is asymptotically stable in the replicator dynamics, but that the strategy $\sigma^* = \left(\frac{1}{3}, \frac{1}{3}, \frac{1}{3}\right)$ is not an ESS. [Hint: consider the strategy $\sigma = (0, \frac{1}{2}, \frac{1}{2})$].

	A	B	C
A	0,0	1,-2	1,1
B	-2,1	0,0	3,1
C	1,1	1,3	0,0

If the derivative of the relative entropy function for a fixed point (taken along solution trajectories) is positive, then the fixed point is unstable. If the derivative is zero, then the fixed point is neither asymptotically stable nor unstable: the evolution of the population is periodic around the fixed point.

Example 9.18

Consider the Rock-Scissors-Paper game. Let x_1 be the proportion of R-players, x_2 be the proportion of S-players, and x_3 be the proportion of P-players. Then

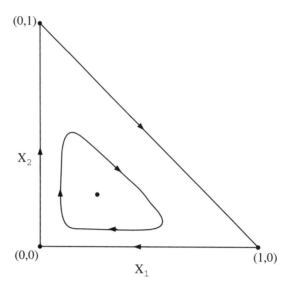

Figure 9.2 A qualitative picture of the replicator dynamics for the Rock-Scissors-Paper game analysed in Example 9.18. We have used the reduced state vector description with $x_3 = 1 - x_1 - x_2$.

the replicator dynamics system is

$$\begin{aligned} \dot{x}_1 &= x_1(x_2 - x_3) \\ \dot{x}_2 &= x_2(x_3 - x_1) \\ \dot{x}_3 &= x_3(x_1 - x_2) \end{aligned}$$

with fixed points $(1, 0, 0)$, $(0, 1, 0)$, $(0, 0, 1)$, and $\left(\frac{1}{3}, \frac{1}{3}, \frac{1}{3}\right)$. It is easy to see by considering the boundaries that the first three points are not stable. For example, consider the invariant line $x_1 = 0$ where, for $0 < x_2, x_3 < 1$, we have $\dot{x}_2 > 0$ and $\dot{x}_3 < 0$. The results from the three invariant lines together imply that there is some kind of oscillatory behaviour about the polymorphic fixed point $\left(\frac{1}{3}, \frac{1}{3}, \frac{1}{3}\right)$: if the fixed point is asymptotically stable, then trajectories will spiral into toward it; if it is unstable, then trajectories will spiral out from it. The third possibility is that solution trajectories form closed loops around the fixed point. That this is, in fact, the case can be confirmed by observing that the time derivative of the relative entropy along solution trajectories of the replicator dynamics is

$$\begin{aligned} \frac{d}{dt}V(\mathbf{x}) &= -\frac{1}{3}(x_2 - x_3) - \frac{1}{3}(x_3 - x_1) - \frac{1}{3}(x_1 - x_2) \\ &= 0. \end{aligned}$$

9.5 Equilibria and Stability

A qualitative picture of the replicator dynamics for the Rock-Scissors-Paper game is shown in Figure 9.2.

Exercise 9.10

Consider a variation of the Rock-Scissors-Paper game in which there is a cost to both players (payoff $= -c$) only if the result is a draw. Show that $\mathbf{x}^* = \left(\frac{1}{3}, \frac{1}{3}, \frac{1}{3}\right)$ is asymptotically stable in the replicator dynamics.

Part IV

Appendixes

A
Constrained Optimisation

Suppose we want to maximise a function of two variables, $f(x,y)$, subject to a constraint $g(x,y) = 0$ that expresses an implicit relation between x and y. In some cases, this implicit relationship may be easily turned into an explicit one of the form $y = h(x)$, and the maximum of the function can be found by differentiating $f(x, h(x))$ with respect to x.

Example A.1

To maximise the function $f(x,y) = x - y^2$ subject to the constraint $x - y = 0$, rewrite the constraint as $y = x$ and differentiate $f(x,x) = x - x^2$. The maximum is then at $(x^*, y^*) = \left(\frac{1}{2}, \frac{1}{2}\right)$.

There is an alternative approach to constrained optimisation: the method of Lagrange multipliers. (As we will show later, this method has the advantage that it can also be applied in cases where the constraint is in the form $g(x,y) \leq 0$ and direct substitution is impossible.) First we combine the function to be maximised, $f(x,y)$, and the function defining the constraint, $g(x,y)$, into a single function called the *Lagrangian*[1]

$$L(x,y) = f(x,y) - \lambda g(x,y).$$

---

[1] As this function is named after the French mathematician Joseph Louis Lagrange (1736–1813), its name is sometimes written as "Lagrangean".

(The, as yet unknown, constant λ is called the *Lagrange multiplier* and will be determined once we have found the maximum we require.) Then we perform an unconstrained maximisation of the Lagrangian. One way of looking at this procedure is to view it as an unconstrained maximisation of the original function $f(x, y)$ with an additional penalty for violating the constraint $g(x, y) = 0$. The following theorem shows that following this procedure does, indeed, produce a maximum of the original function of interest subject to the imposed constraint.

Theorem A.2

If $L(x^*, y^*)$ is an unconstrained maximum of the Lagrangian, then the maximum of the function $f(x, y)$ subject to the constraint $g(x, y) = 0$ occurs at the point (x^*, y^*).

Proof

The maximum of the Lagrangian is found by solving simultaneously the three equations obtained by differentiating with respect to x, y, and λ.

$$\frac{\partial f}{\partial x} - \lambda \frac{\partial g}{\partial x} = 0$$
$$\frac{\partial f}{\partial y} - \lambda \frac{\partial g}{\partial y} = 0$$
$$g(x, y) = 0$$

Note that differentiating by λ will always lead to the constraint being satisfied at the maximum, if one can be found: i.e., $g(x^*, y^*) = 0$. Then

$$\begin{aligned} f(x^*, y^*) &= L(x^*, y^*) + \lambda g(x^*, y^*) \\ &= L(x^*, y^*) \\ &\geq L(x, y) \quad \forall x, y \\ &= f(x, y) \quad \forall x, y \text{ such that } g(x, y) = 0 \end{aligned}$$

□

Example A.3

To maximise the function $f(x, y) = x - y^2$ subject to the constraint $x - y = 0$, introduce the Lagrangian

$$L(x, y) = x - y^2 - \lambda(x - y).$$

A. Constrained Optimisation

Differentiating with respect to x, y, and λ gives the three equations

$$1 - \lambda = 0$$
$$-2y^* + \lambda = 0$$
$$x^* - y^* = 0$$

with the solution $x^* = y^* = \frac{1}{2}$. We also know that $\lambda = 1$ but that is not relevant to the solution of our original problem.

We will now show how this method can be extended to situations in which the constraint is a weak inequality rather than an equality: that is, $g(x, y) \leq 0$. It is, in fact, quite simple: we construct the same Lagrangian $L(x, y) = f(x, y) - \lambda g(x, y)$, but now we also require $\lambda \geq 0$.

Theorem A.4

If $L(x^*, y^*)$ is an unconstrained maximum of the Lagrangian and $\lambda \geq 0$, then the maximum of the function $f(x, y)$ subject to the constraint $g(x, y) \leq 0$ occurs at the point (x^*, y^*).

Proof

Suppose that we have found an unconstrained maximum of the Lagrangian with $\lambda \geq 0$. Then, as before, we have $g(x^*, y^*) = 0$. So

$$\begin{aligned} f(x^*, y^*) &= L(x^*, y^*) \\ &\geq L(x, y) \quad \forall x, y \\ &= f(x, y) - \lambda g(x, y) \\ &\geq f(x, y) \quad \forall x, y \text{ such that } g(x, y) \leq 0 \end{aligned}$$

$\square$

The following example should help to clarify the use of Theorem A.4.

Example A.5

Suppose we wish to maximise the function $f(x, y) = 2x^2 + y^2$ subject to the constraint $x^2 + y^2 \leq 1$. First we put the constraint in the required form: $x^2 + y^2 - 1 \leq 0$. Then we construct the Lagrangian, $L(x, y) = 2x^2 + y^2 - \lambda(x^2 + y^2 - 1)$. The three equations that must be satisfied simultaneously are

$$x(2 - \lambda) = 0$$

$$y(1-\lambda) = 0$$
$$x^2 + y^2 = 1.$$

The solutions of these are (i) $\lambda = 1$, $x = 0$ and $y = \pm 1$ and (ii) $\lambda = 2$, $x = \pm 1$ and $y = 0$. In each case λ is positive, as required. But the points $(x, y) = (0, \pm 1)$ and $(x, y) = (\pm 1, 0)$ are only *extrema* of the Lagrangian and not necessarily maxima. Because $f(0, \pm 1) = 1$ but $f(\pm 1, 0) = 2$ only the points $(x, y) = (\pm 1, 0)$ are maxima of $f(x, y)$.

B
Dynamical Systems

Although it is easy to analyse the behaviour of a one-dimensional dynamical system, such as the replicator dynamics for two-strategy pairwise contest games, it is much more difficult to understand the behaviour of a system in two dimensions or more. In Section 9.3, we introduced a linearisation procedure to provide a connection between the stability of replicator dynamics fixed points and ESSs. Because we can easily understand the behaviour of the full, non-linear system, it is apparent that the picture of the behaviour near to the fixed points is the same whether we consider the full system or its linearised approximation. If this relationship holds true for systems with two or more dimensions[1], then we have some hope of understanding the full system by considering its linearised approximation. With this in mind, we begin by considering linear dynamical systems.

Linear Dynamical Systems

For simplicity, let us consider a two-dimensional linear dynamical system. This can be written as

$$\dot{x}_1 = ax_1 + bx_2$$
$$\dot{x}_2 = cx_1 + dx_2$$

---
[1] It will turn out that this often, but not always, is the case.

where a, b, c, and d are constants. (We have assumed that the unique fixed point of the system occurs at the origin. If it doesn't, we can just shift the coordinate system so that it does.) We can also write this in matrix form as

$$\begin{pmatrix} \dot{x}_1 \\ \dot{x}_2 \end{pmatrix} = \begin{pmatrix} a & b \\ c & d \end{pmatrix} \begin{pmatrix} x_1 \\ x_2 \end{pmatrix}$$

or even more compactly as

$$\dot{\mathbf{x}} = L\mathbf{x} \qquad (\text{B.1})$$

where

$$L = \begin{pmatrix} a & b \\ c & d \end{pmatrix}.$$

If the system happens to have the special form

$$\begin{pmatrix} \dot{z}_1 \\ \dot{z}_2 \end{pmatrix} = \begin{pmatrix} \lambda_1 & 0 \\ 0 & \lambda_1 \end{pmatrix} \begin{pmatrix} z_1 \\ z_2 \end{pmatrix}$$

then the solution is straightforward:

$$\mathbf{z}(t) = \begin{pmatrix} z_1(t) \\ z_2(t) \end{pmatrix} = \begin{pmatrix} z_1(0)e^{\lambda_1 t} \\ z_2(0)e^{\lambda_2 t} \end{pmatrix}. \qquad (\text{B.2})$$

We solve the more general case by finding a transformation of variables $\mathbf{x} \to \mathbf{z}$ such that Equation (B.1) can be written as

$$\dot{\mathbf{z}} = \Lambda \mathbf{z} \qquad (\text{B.3})$$

with

$$\Lambda = \begin{pmatrix} \lambda_1 & 0 \\ 0 & \lambda_1 \end{pmatrix}.$$

Suppose that such a transformation can be achieved through multiplication by a matrix T: $\mathbf{z} = T\mathbf{x}$. Then the solution $\mathbf{x}(t)$ of Equation (B.1) can be found by applying the inverse transformation to Equation (B.2):

$$\mathbf{x}(t) = T^{-1}\mathbf{z}(t).$$

So we need to find out two things: what are the elements of the matrix T^{-1} and what are the constants λ_i? Applying T to Equation (B.1) gives

$$\begin{aligned} T\dot{\mathbf{x}} &= TL\mathbf{x} \\ &= TLT^{-1}T\mathbf{x}. \end{aligned}$$

Comparing this with Equation (B.3) shows that we must have

$$TLT^{-1} = \Lambda$$

B. Dynamical Systems

or
$$LT^{-1} = T^{-1}\Lambda.$$

From this we see that the columns of the matrix T^{-1} must be the eigenvectors of the matrix L and the constants λ_i are the associated eigenvalues.[2]

Now we can find the solution of the general linear dynamical system we began with. Let us write
$$T^{-1} = \begin{pmatrix} u_1 & u_2 \\ v_1 & v_2 \end{pmatrix}.$$

Then the solution is[3]

$$\begin{aligned} \mathbf{x}(t) &= T^{-1}\mathbf{z}(t) \\ &= \begin{pmatrix} u_1 & u_2 \\ v_1 & v_2 \end{pmatrix}\begin{pmatrix} z_1(t) \\ z_2(t) \end{pmatrix} \\ &= \begin{pmatrix} u_1 z_1(t) + u_2 z_2(t) \\ v_1 z_1(t) + v_2 z_2(t) \end{pmatrix} \\ &= \mathbf{v}_1 z_1(t) + \mathbf{v}_2 z_2(t) \\ &= \mathbf{v}_1 z_1(0) e^{\lambda_1 t} + \mathbf{v}_2 z_2(0) e^{\lambda_2 t} \end{aligned}$$

where we have introduced
$$\mathbf{v}_i = \begin{pmatrix} u_i \\ v_i \end{pmatrix}$$

which are the eigenvectors of L. This solution specifies an evolution of the system $\mathbf{x}(t) = (x(t), y(t))$ for each initial state $\mathbf{x}(0) = (x(0), y(0))$. The initial state determines values of the constants $z_i(0)$.

Remark B.1

Nothing in the previous discussion actually depends on the number of equations being 2. So we can immediately find the solution of an n-dimensional linear dynamical system $\dot{\mathbf{x}} = L\mathbf{x}$ (where L is an $n \times n$ matrix). It is

$$\mathbf{x}(t) = \sum_{i=1}^{n} C_i \mathbf{v}_i e^{\lambda_i t}$$

where the $\mathbf{v}_i$ are the eigenvectors and the λ_i are the eigenvalues of the matrix L and the C_i are constants whose values depend on the initial state.

---

[2] If $\mathbf{v}$ is the eigenvector of a matrix L with associated eigenvalue λ, then $L\mathbf{v} = \lambda\mathbf{v}$.
[3] For simplicity, we are ignoring "degenerate cases" in which there is either a single repeated eigenvalue or one or more of the eigenvalues is zero.

Example B.2

Consider the system of equations

$$\dot{x} = x - y$$
$$\dot{y} = y - x$$

which can be written in matrix form with

$$L = \begin{pmatrix} 1 & -1 \\ -1 & 1 \end{pmatrix}.$$

Solving the characteristic equation

$$\begin{aligned} 0 &= \det(L - \lambda I) \\ &= \det \begin{pmatrix} 1 - \lambda & -1 \\ -1 & 1 - \lambda \end{pmatrix} \\ &= (1 - \lambda)^2 - 1 \end{aligned}$$

gives the two eigenvalues as $\lambda_1 = 2$ and $\lambda_2 = 0$. The corresponding eigenvectors are found from the equation $L\mathbf{x} = \lambda \mathbf{x}$. For $\lambda_1 = 2$ we have

$$\begin{pmatrix} 1 & -1 \\ -1 & 1 \end{pmatrix} \begin{pmatrix} x \\ y \end{pmatrix} = 2 \begin{pmatrix} x \\ y \end{pmatrix}$$

which gives $x = y$. Thus, the vector

$$\mathbf{v}_1 = \begin{pmatrix} 1 \\ 1 \end{pmatrix}$$

(or any scalar multiple of it) is an eigenvector corresponding to the eigenvalue $\lambda_1 = 2$. Similarly, an eigenvector for the eigenvalue $\lambda_2 = 0$ is

$$\mathbf{v}_2 = \begin{pmatrix} 1 \\ -1 \end{pmatrix}.$$

Therefore, the solution of the dynamical system is

$$\begin{aligned} x(t) &= z_1(0)e^{2t} + z_2(0) \\ y(t) &= z_1(0)e^{2t} - z_2(0) \end{aligned}$$

where the constants $z_i(0)$ are determined by the initial state of the system, $x(0)$ and $y(0)$.

Remark B.3

In general, both the eigenvalues and the eigenvectors of the matrix L may be complex. However, the constants $z_i(0)$ will also be complex in such a way that the final expressions for $\mathbf{x}(t)$ are always real.

B. Dynamical Systems

We will not be concerned with the exact solution of a linear dynamical system because it will, in general, only be an approximation to the full nonlinear system that we are really interested in. What will be of interest is the qualitative behaviour of the system near the fixed point. The solution is the sum of terms involving $e^{\lambda t}$ where each complex eigenvalue can be written in the form $\lambda = \alpha + i\omega$. Because

$$e^{(\alpha+i\omega)t} = e^{\alpha t}(\cos\omega t + i\sin\omega t)$$

we can make the following observations:

1. If $\alpha < 0$ for all eigenvalues, then $\mathbf{x}(t)$ approaches the fixed point at $t \to \infty$. That is, the fixed point is asymptotically stable.

2. If $\alpha > 0$ for one or more eigenvalues, then $\mathbf{x}(t)$ diverges from the fixed point along the directions of the corresponding eigenvectors. That is, the fixed point is unstable.

3. For $\omega \neq 0$, there is some sort of cyclic behaviour. If the fixed point is stable, then $\mathbf{x}(t)$ spirals in towards the fixed point as $t \to \infty$. If the fixed point is unstable, then $\mathbf{x}(t)$ spirals out from the fixed point. If all eigenvalues are imaginary (i.e., have $\alpha = 0$) then trajectories form ellipses around the fixed point.

Table B.1 shows the classification of the fixed points of a linear dynamical system $\dot{\mathbf{x}} = L\mathbf{x}$ according to the nature of the eigenvalues of the matrix L.

Table B.1 Classification of the fixed points of the linear dynamical system $\dot{\mathbf{x}} = L\mathbf{x}$ according to the real and imaginary parts of the eigenvalues of L. In the case "All $\alpha \neq 0$", some are positive and some negative.

All $\alpha < 0$	$\omega = 0$	Stable node
All $\alpha > 0$	$\omega = 0$	Unstable node
All $\alpha \neq 0$	$\omega = 0$	Saddle point
All $\alpha < 0$	$\omega \neq 0$	Stable spiral
All $\alpha > 0$	$\omega \neq 0$	Unstable spiral
All $\alpha = 0$	$\omega \neq 0$	Centre

Non-linear Dynamical Systems

To obtain a qualitative understanding of a non-linear dynamical system, we follow a 4-step procedure:

1. Find the fixed points of the non-linear system.

2. Derive a linear approximation of the system close to each fixed point.

3. Determine the properties of the fixed point in the linearised system (see Table B.1).

4. Combine this information to produce a sketch of the full, non-linear system.

Example B.4

Consider the two-dimensional dynamical system

$$\dot{x} = x(1-x)(1-2y)$$
$$\dot{y} = y(1-y)(1-2x)$$

defined on the unit square (i.e., $0 \leq x, y \leq 1$). The fixed points (x^*, y^*) of this system are the set of points $\{(0,0)(0,1)(1,0)(1,1)(\frac{1}{2}, \frac{1}{2})\}$. Now we linearise the system close to each of the fixed points in turn.

$(x^*, y^*) = (0, 0)$
Write $\xi = x - x^* = x$ and $\eta = y - y^* = y$ then

$$\dot{\xi} = \xi(1-\xi)(1-2\eta)$$
$$\dot{\eta} = \eta(1-\eta)(1-2\xi).$$

Ignoring non-linear terms (i.e., terms of the form $\xi^n \eta^m$ with $n + m > 1$), we have the linear approximation

$$\dot{\xi} = \xi$$
$$\dot{\eta} = \eta.$$

Clearly this fixed point is an unstable node ($\lambda_1 = \lambda_2 = 1$).

$(x^*, y^*) = (0, 1)$
Write $\xi = x - x^* = x$ and $\eta = y - y^* = y - 1$, then

$$\dot{\xi} = \xi(1-\xi)(1-2(1-\eta))$$
$$\dot{\eta} = -(1+\eta)\eta(1-2\xi).$$

B. Dynamical Systems

Ignoring non-linear terms, we have the linear approximation

$$\dot{\xi} = -\xi$$
$$\dot{\eta} = -\eta.$$

Clearly this fixed point is a stable node ($\lambda_1 = \lambda_2 = -1$).

$(x^*, y^*) = (1, 0)$
Writing $\xi = 1 - x$ and $\eta = y - 0$, we find that this fixed point is an unstable node.

$(x^*, y^*) = (1, 1)$
Writing $\xi = 1 - x$ and $\eta = 1 - y$, we find that this fixed point is a stable node.

$(x^*, y^*) = (\frac{1}{2}, \frac{1}{2})$
Write $\xi = x - \frac{1}{2}$ and $\eta = y - \frac{1}{2}$, then

$$\dot{\xi} = (\frac{1}{2} - \xi)^2(-2\eta)$$
$$\dot{\eta} = (\frac{1}{2} - \eta)^2(-2\xi).$$

Ignoring non-linear terms, we have the linear approximation

$$\dot{\xi} = -\frac{1}{2}\eta$$
$$\dot{\eta} = -\frac{1}{2}\xi.$$

Because the matrix

$$L = \begin{pmatrix} 0 & -\frac{1}{2} \\ -\frac{1}{2} & 0 \end{pmatrix}$$

has eigenvalues $\lambda_1 = \frac{1}{2}$ and $\lambda_2 = -\frac{1}{2}$ with corresponding eigenvectors

$$\mathbf{e}_1 = \begin{pmatrix} 1 \\ -1 \end{pmatrix} \quad \text{and} \quad \mathbf{e}_2 = \begin{pmatrix} 1 \\ 1 \end{pmatrix}.$$

this fixed point is a saddle point with stable direction $x = y$ and unstable direction $x = -y$.

The behaviour of the (linearised) system near each of the fixed points is shown in Figure B.1. Based on this, it seems reasonable that the behaviour of solutions of the full system should (qualitatively) look like that shown in Figure B.2.

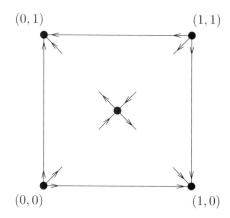

Figure B.1 The behaviour of the linearised system from Example B.4 near each of the fixed points.

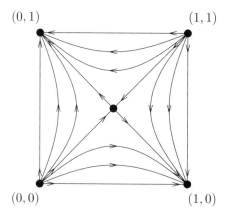

Figure B.2 Solutions of the full system from Example B.4 can be constructed by linking the fixed points in a way indicated by their properties in the linearised system, as shown in Figure B.1.

In the previous example, we went from the picture of a system that was based on a linear approximation near the fixed points to a more complete (although qualitative) picture of the full system. We did this by joining up the fixed points in a way indicated by the properties of the fixed points in the linearised system. This procedure relies on two assumptions. First, we have assumed that the properties of fixed points in the full system are similar to the properties of those fixed points in the linearised approximation. Second, we have assumed that the solution that passes through any given point that is not

B. Dynamical Systems

a fixed point is unique and that the solution changes continuously as the given point is varied. The validity of these assumptions is confirmed by the following two theorems, which we state without proof.

Definition B.5

A fixed point of a dynamical system is called *hyperbolic* if the linearisation of the system near the fixed points has no eigenvalues with a zero real part.

Theorem B.6

The Hartman-Gobman theorem. If a fixed point is hyperbolic, then the topology of the fixed point in the full, non-linear system is the same as the topology of the fixed point in the linearised system.

Remark B.7

The Hartman-Grobman theorem justifies the use of the linearisation approach to discovering the properties of fixed points in a dynamical system in most cases. The exceptions are situations where one or more eigenvalues are purely imaginary. In such cases, the stability (or otherwise) of the fixed point must be determined by other means.

Theorem B.8

Consider a dynamical system $\dot{\mathbf{x}} = \mathbf{f}(\mathbf{x})$. If the vector field $\mathbf{f}$ has continuous first derivatives at a point $\mathbf{x}(0)$, then (i) there is a unique solution $\mathbf{x}(t)$ that passes through $\mathbf{x}(0)$ and (ii) the solution $\mathbf{x}(t)$ changes smoothly as the point $\mathbf{x}(0)$ is varied.

Remark B.9

In the replicator dynamics for pairwise contest games, the functions $\mathbf{f}$ are polynomials and are, therefore, always suitably well-behaved. In particular, this theorem means that trajectories in the replicator dynamics cannot cross.

As we have already remarked, the stability (or otherwise) of a non-hyperbolic fixed point cannot be determined by linearisation. The linearised system has eigenvalues with real parts that are zero, and the stability of the system is determined by higher order terms in expansion about the fixed point. In such cases, we can make use of the *direct Lyapounov method* (named after the Russian mathematician who devised it).

Theorem B.10

Let $\dot{\mathbf{x}} = \mathbf{f}(\mathbf{x})$ be a dynamical system with a fixed point at $\mathbf{x}^*$. If we can find a scalar function $V(\mathbf{x})$, defined for allowable states of the system close to $\mathbf{x}^*$, such that[4]

1. $V(\mathbf{x}^*) = 0$
2. $V(\mathbf{x}) > 0$ for $\mathbf{x} \neq \mathbf{x}^*$
3. $\frac{dV}{dt} < 0$ for $\mathbf{x} \neq \mathbf{x}^*$

then the fixed point $\mathbf{x}^*$ is asymptotically stable.

Proof

The third condition implies that V is strictly decreasing along solution trajectories of the dynamical system. Because $V(\mathbf{x}) \geq 0$ with equality only for $\mathbf{x} = \mathbf{x}^*$, this implies that $\lim_{t \to \infty} V = 0$ and hence $\lim_{t \to \infty} \mathbf{x}(t) = \mathbf{x}^*$. □

Remark B.11

The drawback of the direct Lyapounov method is that there is no general procedure (apart from trial and error) for constructing a Lyapounov function. For the replicator dynamics, however, a class of functions known as *relative entropy functions* seems to work well in many cases.

Example B.12

Consider the dynamical system

$$\begin{aligned} \dot{x} &= -y \\ \dot{y} &= x - x^2 y. \end{aligned}$$

In the linearised system

$$\begin{aligned} \dot{\xi} &= -\eta \\ \dot{\eta} &= \xi. \end{aligned}$$

---

[4] A function with these properties is called a strict Lyapounov function.

the fixed point at the origin is a centre (both eigenvalues are purely imaginary). Because

$$\frac{d}{dt}(\xi^2 + \eta^2) = 2\xi\dot{\xi} + 2\eta\dot{\eta}$$
$$= 0$$

trajectories (in the linear system) form concentric circles about the origin: $\xi(t)^2 + \eta(t)^2 = $ constant. This tells us nothing about the full non-linear system, but it does give the hint that we should try $V = x^2 + y^2$ as a Lyapounov function. Clearly this function satisfies the first two conditions, and because

$$\begin{aligned}\frac{dV}{dt} &= 2x\dot{x} + 2y\dot{y} \\ &= 2x(-y) + 2y(x - x^2 y) \\ &= -2x^2 y^2 \\ &< 0 \quad \text{for } \mathbf{x} \neq \mathbf{x}^*\end{aligned}$$

it satisfies the third as well. The origin is, therefore, an asymptotically stable fixed point.

Solutions

Chapter 1

1.1 Value of $F(n)$ for various n is

n	0	1	2	3	4	5+
$F(n)$	1	9/2	6	11/2	3	<0

So $n^* = 2$ and $F(n^*) = 6$.

1.2 Because $g(0) = 0$ and $g(b/2) < 0$ for $ab < 4c$

$$x^* = \begin{cases} \frac{b}{2} & \text{if } ab > 4c \\ 0 & \text{otherwise} \end{cases}.$$

1.3 (a) $x^* = 3$ ($f'(3) = 0$ & $f''(3) < 0$).
(b) $x^* = 2$ ($f'(x) > 0$ for $x \in [1,2]$).
(c) $x^* = 3$ ($f''(x) > 0$).

1.4 The payoff function is the return on the investment, because the two accounts are otherwise identical. Let us first assume that the initial capital is included. Then $\pi(a_1) = £1000 \times 1.06 = £1060$ and $\pi(a_2) = £1000 \times (1.03)^2 = £1060.90$. So the investor should choose the second account, which pays 3% at six-month intervals. The result will be the same if the initial sum is not included (it is an affine transformation).

1.5 (a) Income $= qP(q)$, which is maximised at $\hat{q} = \frac{1}{2}q_0$.
(b) Profit $= q(P(q) - c)$, which is maximised at

$$q^* = \frac{q_0}{2}\left(1 - \frac{c}{P_0}\right)$$

1.6 Profit $= (p - c_1)T(p) - c_0$, which is maximised at

$$p^* = \frac{p_0 + c_1}{2}.$$

The maximum profit is only positive if $T(p^* - c_1)(p^*) - c_0 > 0$. This condition is satisfied if

$$T_0\left(1 - \frac{c_1}{p_0}\right)^2 > 4\frac{c_0}{p_0}.$$

If this condition is not satisfied, then the factory should not be built.

1.7 (a) $\pi(a_1) = \pi(a_2) = 1.5$ and $\pi(a_3) = 1$. So $a^* = a_1$ or a_2 but not a_3.
(b) $\pi(a_1) = 0.75$, $\pi(a_2) = 2.25$ and $\pi(a_3) = 1$. So $a^* = a_2$.

1.8 The expected profit is

$$\begin{aligned}\pi(u) &= \int_0^u (px - cu)f(x)\,dx + \int_u^\infty (px - cu - k(x-u))f(x)\,dx \\ &= \int_0^\infty (px - cu)f(x)\,dx - k\int_u^\infty (x - u)f(x)\,dx \\ &= pd - cu - k\int_u^\infty xf(x)\,dx + ku\int_u^\infty f(x)\,dx.\end{aligned}$$

Using

$$\int_u^\infty xf(x)\,dx = (u+d)e^{-u/d} \quad \text{and} \quad \int_u^\infty f(x)\,dx = e^{-u/d}$$

we obtain

$$\pi(u) = pd - cu - kde^{-u/d}.$$

Differentiating with respect to u and setting the result equal to zero, we find

$$u^* = d\ln\left(\frac{k}{c}\right).$$

1.9 Because $w \sim N(\mu, \sigma^2)$, the expected utility is

$$\mathbb{E}(u(w)) = 1 - \frac{1}{\sqrt{2\pi}\sigma}\int_{-\infty}^\infty e^{-kw}e^{-(w-\mu)^2/2\sigma^2}.$$

By completing the square, we find

$$\frac{(w-\mu)^2}{2\sigma^2} + kw = \frac{w - \mu + k\sigma^2}{2\sigma^2} + k\mu - \frac{k^2\sigma^2}{2}$$

and

$$\mathbb{E}(u(w)) = 1 - \exp\left[-\left(k\mu - \frac{k^2\sigma^2}{2}\right)\right].$$

So

$$\begin{aligned}\operatorname{argmax} \mathbb{E}(u(w)) &= \operatorname{argmax}\left(k\mu - \frac{k^2\sigma^2}{2}\right) \\ &= \operatorname{argmax}\left(\mu - \frac{k}{2}\sigma^2\right).\end{aligned}$$

Solutions

1.10 The expected utility is
$$\pi = \mathbb{E}(w) - k\mathbb{E}(w^2)$$
$$= \mathbb{E}(w) - k\left(\mathbb{E}(w)\right)^2 - k\mathrm{Var}(w).$$

From example (1.24) we have
$$\mathbb{E}(w) = r + a(\mu - r) \quad \text{and} \quad \mathrm{Var}(w) = a^2\sigma^2$$

so
$$\pi(a) = r - kr^2 + a(\mu - r) - 2akr(\mu - r) - ka^2(\mu - r)^2 - ka^2\sigma^2.$$

This payoff has a maximum at
$$\hat{a} = \frac{(\mu - r)(1 - 2kr)}{2k(\mu - r)^2 + 2k\sigma^2}.$$

So
$$a^* = \begin{cases} 0 & \text{if } \hat{a} < 0 \\ \hat{a} & \text{if } 0 < \hat{a} < 1 \\ 1 & \hat{a} > 1 \end{cases}.$$

1.11 Expected number of surviving offspring is $nH(n)$.

n	0	1	2	3	4+
$nH(n)$	0	0.9	1.2	0.3	<0

So $nH(n)$ has a maximum at $n^* = 2$

1.12 Maximising the expected number of offspring is equivalent to maximising the adult's probability of survival. The probability of survival, if the bird "chooses" site i, is
$$S(i) = (1 - \lambda_i)\left(P_i M_h + (1 - P_i)M_l\right).$$

So
$$S(1) = 0.656 \quad S(2) = 0.666 \quad S(3) = 0.627$$

which gives the optimal patch choice as $i^* = 2$.

1.13 The payoff for a general behaviour β is
$$\sum_{a \in \mathbf{A}} p(a)\pi(a) = \sum_{a \in \mathbf{A}} p(a) \sum_{x \in \mathbf{X}} P(X = x)\pi(a|x)$$
$$= \sum_{x \in \mathbf{X}} P(X = x) \sum_{a \in \mathbf{A}} p(a)\pi(a|x)$$
$$= \sum_{x \in \mathbf{X}} P(X = x)\pi(\beta|x).$$

1.14 Because $\pi(a_1) = \pi(a_3) = 5$ and $\pi(a_2) = 3\frac{3}{4}$, optimal randomising behaviours have support $\mathbf{A}^* = \{a_1, a_2\}$ with $p(a_1) = p$ and $p(a_3) = 1 - p$ $(0 < p < 1)$. Using either a_1 or a_3 with probability 1 is also an optimal behaviour.

Chapter 2

2.1 (a) The decision tree is

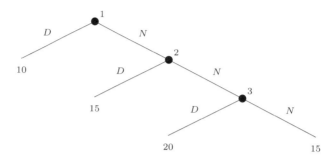

The pure-strategy set is

$$\mathbf{S} = \{NNN, NND, NDN, NDD, DNN, DND, DDN, DDD\}$$

and the optimal strategy is NND, which gives a payoff of 20 cents.
(b) If play has reached the last choice, then it is optimal to choose the dime rather than the nickel. At any other point, it is optimal to choose the nickel because, for example, choosing the nickel now and the dime next time gives a total future payoff of 15 cents compared to the 10 cents gained by choosing the dime now.
(c) At any decision point, the probability that the game will continue for another n choices is $\sum_{k=1}^{n} p^k$, so the expected future number of decisions is

$$\sum_{k=1}^{\infty} p^k = \frac{p}{1-p}$$

In this sense, all decision points are the same, so if it is optimal to choose an action now, it will be optimal to choose the same action in the future (i.e., the optimal strategy is stationary). The expected future payoff for choosing the nickel is

$$5 + \frac{5p}{1-p}$$

whereas the payoff for choosing the dime is just 10 (because the game then stops). So it is optimal to choose the nickel every time if $p > \frac{1}{2}$.

2.2 The payoff for the behavioural strategy is

$$\pi(\beta) = \frac{1}{2} \times 15 + \frac{1}{2} \times 10 = 12.5$$

The payoff for the mixed strategies is

$$\pi(\sigma) = \frac{1}{2} \times 15 + \left(\frac{1}{2} - x\right) \times 10 + x \times 10 = 12.5$$

2.3 (a) Because all decision points are reached with positive probability by playing the mixed strategy, the behavioural equivalent $\beta = \left(\left(\frac{1}{2},\frac{1}{2}\right),(1,0),(0,1)\right)$ is unique.

(b) Because decision point 3 is not reached by playing the mixed strategy, any choice at that point gives a behavioural equivalent. So the equivalent behavioural strategies are all of the form

$$\beta = \left((1,0),\left(\frac{2}{3},\frac{1}{3}\right),(x,1-x)\right)$$

with $x \in [0,1]$.

2.4 (a) Using the notation C for care, D for desert, and H, M, and L for the choice of the high-quality, medium-quality or low-quality male, the decision tree is

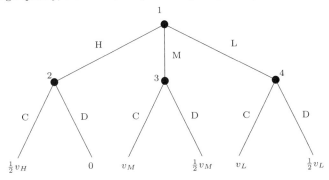

(b) The female should always care and should choose her mate according to the rule:

$$\text{Choose} \begin{cases} H & \text{if } \frac{1}{2}v_H > v_M \\ M & \text{if } \frac{1}{2}v_H < v_M \\ L & \text{never} \end{cases}.$$

Chapter 3

3.1 (a) The payoff is

$$\pi(c_0,c_1,c_2) = \ln(c_0) + \ln(c_1) + \ln(c_2)$$

and the constraint is

$$c_0 + \frac{c_1}{r} + \frac{c_2}{r^2} \le x_0$$

which gives the solution

$$c_{t+1} = rc_t \quad \text{with} \quad c_0 = \frac{x_0}{3}.$$

(b) By backward recursion the solution is

$$c_2^* = x_2 \qquad c_1^* = \frac{1}{2}x_1 \qquad c_0^* = \frac{x_0}{3}.$$

Substituting in the state equation, we find

$$c_1^* = \frac{1}{2}r(x_0 - c_0) = \frac{1}{2}r\frac{2}{3}x_0 = rc_0^*$$

$$c_2^* = r(x_1 - c_1) = r\frac{1}{2}x_1 = rc_1^*$$

3.2 The correspondence is given in the following table.

General description	Example 3.4
$r_T(x_T)$	0
$r_t(x_t, a_t)$	$\ln(c_t)$
$p(x'\|x,a)$	$p(x' = r(x-c)\|x,c) = 1$
$a(x_t, t)$	$c_t(x_t)$
s	$(c_0(x_0), c_1(x_1))$
X	$[0, \infty]$
A(x,t)	$[0, x_t]$

3.3 Beginning in state x, the process proceeds as follows

```
t = 0           t = 1            t = 2            t = 3
         p=1             p=1/2            p=1/2
(x,a) ────────→ (y,b) ─────────→ (y,b) ─────────→ y
                      ╲                 ╲
                   p=1/2              p=1/2
                        ╲                 ╲
                         z                 z
```

and the expected payoff is $2 + 10 + \frac{1}{2}10 = 17$. Beginning in state y, the process proceeds as follows

```
t = 0            t = 1             t = 2             t = 3
        p=1/2             p=1/2              p=1/2
(y,b) ─────────→ (y,b) ─────────→ (y,b) ─────────→ y
       ╲                 ╲                  ╲
     p=1/2             p=1/2              p=1/2
         ╲                 ╲                  ╲
          z                 z                  z
```

and the expected payoff is $10 + \frac{1}{2}10 + \frac{1}{4}10 = 17.5$.

3.4 In state z, there is no choice to be made and $\pi_t^*(z) = 0$, $\forall t$. The absence of a terminal reward r_T also gives us $\pi_3^*(x) = \pi_3^*(y) = 0$.
At time $t = 2$ in state x, we have

$$\pi_2(x|a) = 1 + \pi_3^*(y) = 1$$
$$\pi_2(x|b) = 2 + \pi_3^*(x) = 2$$

so $a^*(x,2) = b$ and $\pi_2^*(x) = 2$. In state y, we have

$$\pi_2(y|a) = 3 + \pi_3^*(x) = 3$$
$$\pi_2(y|b) = 5 + \frac{1}{4}\pi_3^*(y) + \frac{3}{4}\pi_3^*(z) = 5$$

so $a^*(y,2) = b$ and $\pi_2^*(y) = 5$.
At time $t = 1$ in state x, we have

$$\pi_1(x|a) = 1 + \pi_2^*(y) = 6$$
$$\pi_1(x|b) = 2 + \pi_2^*(x) = 4$$

so $a^*(x,1) = a$ and $\pi_1^*(x) = 6$. In state y, we have

$$\pi_1(y|a) = 3 + \pi_2^*(x) = 5$$
$$\pi_1(y|b) = 5 + \frac{1}{4}\pi_2^*(y) + \frac{3}{4}\pi_2^*(z) = 6\frac{1}{4}$$

so $a^*(y,2) = b$ and $\pi_1^*(y) = 6\frac{1}{4}$.
At time $t = 0$ in state x, we have

$$\pi_0(x|a) = 1 + \pi_1^*(y) = 7\frac{1}{4}$$
$$\pi_0(x|b) = 2 + \pi_1^*(x) = 8$$

so $a^*(x,0) = b$ and $\pi_0^*(x) = 8$. In state y, we have

$$\pi_0(y|a) = 3 + \pi_1^*(x) = 9$$
$$\pi_0(y|b) = 5 + \frac{1}{4}\pi_1^*(y) + \frac{3}{4}\pi_1^*(z) = \frac{105}{16} = 6.5625$$

so $a^*(y,2) = a$ and $\pi_0^*(y) = 9$.
The solution of the problem is, therefore, the optimal strategy

$$s^* = \begin{matrix} x \\ y \end{matrix} \begin{pmatrix} t=0 & t=1 & t=2 \\ b & a & b \\ a & b & b \end{pmatrix}$$

where we have ignored state z because it is irrelevant whether we say that the action in state z is a or b. The payoff is 8 if the process starts in state x or 9 if the process starts in state y.

3.5 The payoffs are

$$\pi(s_1) = \sum_{t=0}^{\infty} \delta^t = \frac{1}{1-\delta}$$

$$\pi(s_2) = \sum_{t=0}^{\infty} 2\delta^t = \frac{2}{1-\delta}$$

so s_2 is better than s_1 as we thought it should be.

3.6 Following the strategy $s = \{a(x) = a, a(y) = a\}$ gives

$$\pi(x|s) = 2 + \frac{1}{2}\pi(y|s)$$

$$\pi(y|s) = \frac{1}{2}\pi(z|s)$$

$$\pi(z|s) = 6 + \frac{1}{2}\pi(x|s) .$$

These can be solved (find $\pi(z|s)$ first) to give $\pi(x|s) = \pi(y|s) = 4$ and $\pi(z|s) = 8$. Swapping to action b in state x yields a payoff of $1 + \frac{1}{2}\pi(x|s) = 3$. Swapping to action b in state y yields a payoff of $1 + \frac{1}{2}\pi(x|s) = 3$. From this we can conclude that the strategy we guessed is optimal.

3.7 The optimal strategy is the same as the one found in Example 3.19.

Chapter 4

4.1 One possible payoff table is

		Scarpia	
		Fake execution	Real execution
Tosca	Kill	$\pi_T(K,F), \pi_S(K,F)$	$\pi_T(K,R), \pi_S(K,R)$
	Sleep	$\pi_T(S,F), \pi_S(S,F)$	$\pi_T(S,R), \pi_S(S,R)$

Because Tosca (presumably) prefers to keep her honour in any case, we have $\pi_T(K,F) > \pi_T(S,F)$ and $\pi_T(K,R) > \pi_T(S,R)$. Because Scarpia (presumably) prefers to do his duty in any case, we have $\pi_S(K,R) > \pi_S(K,F)$ and $\pi_S(S,R) > \pi_S(S,F)$. These conditions mean the game inevitably has the stated outcome.

4.2 (a) Eliminate D; then eliminate L, leaving (U, R). (b) Eliminate D (it is dominated by the mixed strategy of playing U and C each with probability $\frac{1}{2}$); then eliminate R; then eliminate U, leaving (C, L). Alternatively, eliminate R, then eliminate U and D, leaving (C, L).

4.3 The pair (D, L) is a Nash equilibrium because

$$\pi_1(\sigma_1, L) = 10p + 10(1-p) = 10 = \pi_1(D, L)$$

and

$$\pi_2(D, \sigma_2) = p - (1 - p - q) \le p \le \pi_2(D, L) .$$

Similarly, the pair (U, M) is a Nash equilibrium because

$$\pi_1(\sigma_1, M) = 5 = \pi_1(U, M)$$

and

$$\pi_2(U, \sigma_2) = q - 2(1 - p - q) \le q \le \pi_2(U, M) .$$

4.4 (a) Let $\sigma_1 = (p, 1-p)$ and $\sigma_2 = (q, 1-q)$, then
$$\pi_1(\sigma_1, \sigma_2) = 1 + q + p(1+q).$$
So the best response for player 1 is $\hat{\sigma}_1 = (1,0)$ (i.e., use U) whatever player 2 does. Similarly
$$\pi_2(\sigma_1, \sigma_2) = 1 + q + p.$$
So player 2's best response is always $\hat{\sigma}_2 = (1,0)$ (i.e., use L). The unique Nash equilibrium is, therefore, (U, L).

(b) Let $\sigma_1 = (p, 1-p)$ and $\sigma_2 = (q, 1-q)$, then $\pi_1(\sigma_1, \sigma_2) = q + p(2-3q)$ so the best responses for P_1 are

$$\hat{\sigma}_1 = \begin{cases} (0,1) & \text{if } q > \frac{2}{3} \\ (1,0) & \text{if } q < \frac{2}{3} \\ (x, 1-x) \text{ with } x \in [0,1] & \text{if } q = \frac{2}{3}. \end{cases}$$

Similarly, $\pi_2(\sigma_1, \sigma_2) = p + q(2-3p)$ so the best responses for P_2 are

$$\hat{\sigma}_2 = \begin{cases} (0,1) & \text{if } p > \frac{2}{3} \\ (1,0) & \text{if } p < \frac{2}{3} \\ (y, 1-y) \text{ with } y \in [0,1] & \text{if } p = \frac{2}{3}. \end{cases}$$

So the complete set of Nash equilibria is (M, R), (F, W) and (σ_1^*, σ_2^*) with $\sigma_1^* = \sigma_2^* = \left(\frac{2}{3}, \frac{1}{3}\right)$.

4.5 We underline the best responses for each player.

		P_2		
		L	M	R
P_1	U	4, 3	2, $\underline{7}$	$\underline{0}$, 4
	D	$\underline{5}, \underline{5}$	$\underline{5}$, -1	$-4, -2$

So the unique pure-strategy Nash equilibrium is (D, L).

4.6 Let $\mathbf{K}$ be the set of integers from 0 to 1000, i.e., $\mathbf{K} = \{0, 1, 2, ..., 999, 1000\}$. The best responses for $s_i \in \mathbf{K}$ are
$$\hat{s}_1 = 1000 - s_2$$
$$\hat{s}_2 = 1000 - s_1.$$
So any pair of sums of money $s_1^* \in \mathbf{K}$ and $s_2^* \in \mathbf{K}$ with $s_1^* + s_2^* = 1000$ is a Nash equilibrium. (There is also an infinite number of Nash equilibria where both sons choose an amount of money greater than £1000!)

4.7 (a) The payoff table is

	R	S	P
R	0, 0	1, -1	-1, 1
S	-1, 1	0, 0	1, -1
P	1, -1	-1, 1	0, 0

(b) There are clearly no pure-strategy Nash equilibria. Find a mixed-strategy equilibrium using the Equality of Payoffs theorem. Let $\sigma_2 = (r, s, 1 - r - s)$, then

$$\pi_1(R, \sigma_2) = \pi_1(S, \sigma_2) = \pi_1(P, \sigma_2)$$
$$2s + r - 1 = 1 - s - 2r = r - s.$$

which can be solved to give $r = s = \frac{1}{3}$. By following the analogous procedure for player 2, we find the unique Nash equilibrium is (σ^*, σ^*) with $\sigma^* = \left(\frac{1}{3}, \frac{1}{3}, \frac{1}{3}\right)$.

4.8 (i) If $a \geq c$, then (A, A) is a symmetric Nash equilibrium. (ii) If $d \geq b$, then (B, B) is a symmetric Nash equilibrium. (iii) If $a < c$ and $d < b$, then there is no symmetric pure strategy Nash equilibrium, so we look for a mixed strategy Nash equilibrium using the Equality of Payoffs theorem. Let $\sigma_1^* = (p^*, 1 - p^*)$ and $\sigma_2^* = (q^*, 1 - q^*)$. Then

$$\pi_1(A, \sigma_2^*) = \pi_1(B, \sigma_2^*)$$
$$\iff aq^* + b(1 - q^*) = cq^* + d(1 - q^*)$$
$$\iff q^* = \frac{(b - d)}{(c - a) + (b - d)}$$

and

$$\pi_2(\sigma_1^*, A) = \pi_2(\sigma_1^*, B)$$
$$\iff ap^* + b(1 - p^*) = cp^* + d(1 - p^*)$$
$$\iff p^* = \frac{(b - d)}{(c - a) + (b - d)}.$$

We have $0 < p^* = q^* < 1$ as required for a symmetric mixed strategy Nash equilibrium.

4.9 The Nash equilibria for both games are (U, L), (D, R) and $\left(\left(\frac{2}{3}, \frac{1}{3}\right), \left(\frac{1}{3}, \frac{2}{3}\right)\right)$.

4.10 (a) Let $\sigma_1 = (p, 1-p)$ and $\sigma_2 = (q, 1-q)$. Then $\pi_1(\sigma_1, \sigma_2) = 6q + 5p(1-q)$, so

$$\hat{\sigma}_1 = \begin{cases} (1, 0) & \forall q < 1 \\ (x, 1 - x) \text{ with } x \in [0, 1] & \text{for } q = 1 \end{cases}.$$

Now $\pi_2(\sigma_1, \sigma_2) = 3p + q(1 - 4p)$, so

$$\hat{\sigma}_2 = \begin{cases} (1, 0) & \forall p < \frac{1}{4} \\ (0, 1) & \forall p > \frac{1}{4} \\ (y, 1 - y) \text{ with } y \in [0, 1] & \text{for } p = \frac{1}{4} \end{cases}.$$

Therefore, the Nash equilibria are

$$(\sigma_1^*, \sigma_2^*) = \begin{cases} ((x, 1 - x), C) & \text{with } x \in [0, \frac{1}{4}] \\ (B, D) & \end{cases}.$$

(b) Let $\sigma_1 = (p, 1-p)$ and $\sigma_2 = (q, r, 1-q-r)$. Then $\pi_1(\sigma_1, \sigma_2) = 2 + 3q - 4r + pr$, so

$$\hat{\sigma}_1 = \begin{cases} (1, 0) & \text{if } r > 0 \text{ and } \forall q \in [0, 1) \\ (x, 1 - x) \text{ with } x \in [0, 1] & \text{if } r = 0 \text{ and } \forall q \in [0, 1] \end{cases}.$$

Now $\pi_2(\sigma_1, \sigma_2) = 3(1-p) + rp$, so

$$\hat{\sigma}_2 = \begin{cases} (0,1,0) & \text{if } p > 0 \\ (y, z, 1-y-z) \text{ with } y, z, y+z \in [0,1] & \text{if } p = 0 \end{cases}.$$

Therefore, the Nash equilibria are

$$(\sigma_1^*, \sigma_2^*) = \begin{cases} (J, (y, 0, 1-y)) & \text{with } y \in [0,1] \\ (G, F) \end{cases}.$$

The best responses for the first game are shown in the figure below. The best responses for player 1 are shown by a solid line and those for player 2 by a dotted line. Where they meet are the Nash equilibria (indicated by the circle and the thick line)

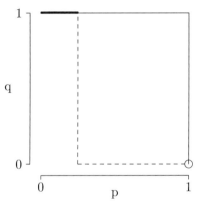

4.11 Let $p = P(\text{player 1 plays } A)$ and $q = P(\text{player 2 plays } C)$, then

$$\pi_1(\sigma_1, \sigma_2) = (2-q) + p(\lambda q - 1)$$
$$\pi_2(\sigma_1, \sigma_2) = (2-p) + q(\lambda p - 1).$$

For $\lambda < 1$, the best responses are $\hat{p} = \hat{q} = 0$ so (D, R) is the unique Nash equilibrium. For $\lambda \geq 1$, the best responses are

$$\hat{p} = \begin{cases} 1 & \text{if } q < \frac{1}{\lambda} \\ 0 & \text{if } q > \frac{1}{\lambda} \\ \text{any } x \in [0,1] & \text{if } q = \frac{1}{\lambda} \end{cases}.$$

and

$$\hat{q} = \begin{cases} 1 & \text{if } p < \frac{1}{\lambda} \\ 0 & \text{if } p > \frac{1}{\lambda} \\ \text{any } y \in [0,1] & \text{if } p = \frac{1}{\lambda} \end{cases}.$$

The Nash equilibria are (U, L), (D, R), and (σ_1^*, σ_2^*) with

$$\sigma_1^* = \sigma_2^* = \left(\frac{1}{\lambda}, \frac{\lambda - 1}{\lambda}\right).$$

As $\lambda \to 1$, $(\sigma_1^*, \sigma_2^*) \to (U, L)$.
For $\lambda \neq 1$, the game is generic and the number of equilibria is odd. For $\lambda = 1$, the game is non-generic and the number of equilibria is even.

4.12 The payoff table is

	A	K	Q
A	0, 0	5, −5	−5, 5
K	−5, 5	0, 0	5, −5
Q	5, −5	−5, 5	0, 0

This game is just an affine transformation of "Rock-Scissors-Paper" so the unique equilibrium is $(\sigma^*, \sigma^*]$ with $\sigma^* = \left(\frac{1}{3}, \frac{1}{3}, \frac{1}{3}\right)$ (see problem 4.7).

4.13 Take each of the sixteen possible cases, one-by-one.

	Sign of			Pure-strategy	Mixed-strategy
$(d-b)$	$(u-c)$	$(d-c)$	$(a-b)$	equilibrium	equilibrium
+	+	+	+	None	p^*, q^*
+	+	+	−	(U, L)	None
+	+	−	+	(D, R)	None
+	+	−	−	Inconsistent	
+	−	+	+	(D, L)	None
+	−	+	−	(D, L)	None
+	−	−	+	(D, R)	None
+	−	−	−	(D, R)	None
−	+	+	+	(U, R)	None
−	+	+	−	(U, L)	None
−	+	−	+	(U, R)	None
−	+	−	−	(U, L)	None
−	−	+	+	Inconsistent	
−	−	+	−	(D, L)	None
−	−	−	+	(U, R)	None
−	−	−	−	None	p^*, q^*

Where

$$p^* = \frac{(d-c)}{(d-c)+(a-b)} \quad \text{and} \quad q^* = \frac{(d-b)}{(d-b)+(a-c)}.$$

The two inconsistent cases arise because $d-b > 0$ and $a-c > 0$ imply that $a-b-c+d > 0$, whereas $d-c < 0$ and $a-b < 0$ imply $a-b-c+d < 0$ (and vice versa).

4.14 The game can be represented by the pair of payoff tables shown below.

	P_3		
	L	A	B
P_1 U	1,1,0	2,2,3
P_1 D	2,2,3	3,3,0

	P_3		
	R	A	B
P_1 U	-1,-1,2	2,0,2
P_1 D	0,2,2	1,1,2

By inspection, there are no pure strategy equilibria. Let player 1 choose U with probability p, let player 2 choose L with probability q, and let player 3 choose A with probability r. The three equations that must be satisfied for mixed strategies are

$$\pi_1(U, \sigma_2, \sigma_3) = \pi_1(D, \sigma_2, \sigma_3)$$
$$\pi_2(\sigma_1, L, \sigma_3) = \pi_2(\sigma_1, R, \sigma_3)$$
$$\pi_3(\sigma_1, \sigma_2, A) = \pi_3(\sigma_1, \sigma_2, B).$$

These yield the following three conditions for p, q, and r:

$$2(q + r - qr) - 1 = 0$$
$$2(p + r - pr) - 1 = 0$$
$$3(p + q - 2pq) = 2.$$

The first two conditions tell us that $p = q$. Because the third condition has no real solutions if $p = q$, player 3 cannot employ a mixed strategy. Suppose $r = 1$, then either of the first two conditions produces the contradiction $1 = 0$. If $r = 0$, then we deduce that $p = q = \frac{1}{2}$.

Chapter 5

5.1 The equilibrium can be written unambiguously as (AEE, CR).

5.2 The second player's strategies are triples XYZ meaning "play X after L, Y after M and Z after R". The backward induction solution is (L, BBB) and the payoff table is

		P_2						
	AAA	AAB	ABA	ABB	BAA	BAB	BBA	BBB
P_1 L	0,0	0,0	0,0	0,0	**6,2**	**6,2**	**6,2**	**6,2**
P_1 M	1,3	1,3	5,4	**5,4**	1,3	1,3	5,4	5,4
P_1 R	6,2	**1,3**	6,2	1,3	6,2	1,3	6,2	1,3

where the pure strategy Nash equilibria are shown in bold type.

5.3 The game has two trees as follows

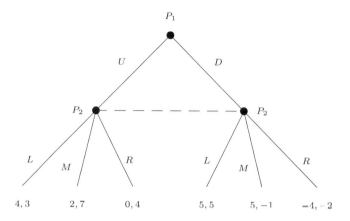

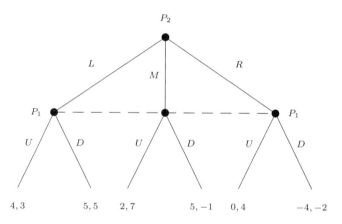

No backward induction solutions are possible because the player that decides "second" does not know what the other player "has" done.

5.4 Based on the version in Example 4.2, the game tree is

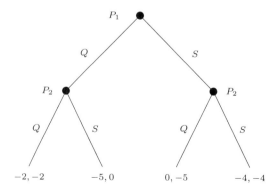

Both players should "squeal" just as they do in the static game.

5.5 The strategic form is

		P_2			
		LH	LT	RH	RT
	AH	1, −1	−1, 1	1, −1	−1, 1
P_1	AT	−1, 1	1, −1	−1, 1	1, −1
	BH	3, 1	3, 1	−1, 2	−1, 2
	BT	3, 1	3, 1	−1, 2	−1, 2

The mixed strategies $\sigma_1^* = \frac{1}{2}AH + \frac{1}{2}AT$ and $\sigma_2^* = \frac{1}{2}RH + \frac{1}{2}RT$ give $\pi_1(\sigma_1^*, \sigma_2^*) = \pi_2(\sigma_1^*, \sigma_2^*) = 0$. Because $\pi_1(AH, \sigma_2^*) = \pi_1(AT, \sigma_2^*) = 0$ and $\pi_1(BH, \sigma_2^*) = \pi_1(BT, \sigma_2^*) = -1$, we have $\pi_1(\sigma_1^*, \sigma_2^*) \geq \pi_1(\sigma_1, \sigma_2^*) \ \forall \sigma_1 \in \Sigma_1$. Because $\pi_2(\sigma_1^*, s_2) = 0 \ \forall s_2 \in \mathbf{S}_2$, we have $\pi_2(\sigma_1^*, \sigma_2^*) \geq \pi_2(\sigma_1^*, \sigma_2) \ \forall \sigma_2 \in \Sigma_2$. Hence (σ_1^*, σ_2^*) is a Nash equilibrium.

5.6 The Newcomer's pure strategies are to "enter the market" (E) or to "stay out" (S). The Incumbent's pure strategies are to "engage in a price war" (W) or to "accept the competition" (A). The game tree is

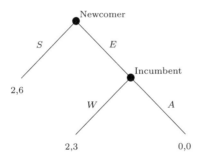

and (E, A) is the unique behavioural strategy equilibrium.
The strategic form is

	W	A
S	2, 6	2, 6
E	1, 1	3, 3

Let $p = P(\text{Newcomer plays } S)$ and $q = P(\text{Incumbent plays } W)$, then

$$\pi_N(\sigma_N, \sigma_I) = 3 - 2q + p(2q - 1)$$

and

$$\pi_I(\sigma_N, \sigma_I) = 3 + 3p + 2q(p - 1)$$

So the best responses for the Newcomer are

$$\hat{\sigma}_N = \begin{cases} (0,1) & \text{if } q < \frac{1}{2} \\ (1,0) & \text{if } q > \frac{1}{2} \\ (x, 1-x) \text{ with } x \in [0,1] & \text{if } q = \frac{1}{2} \end{cases}$$

and the best responses for the Incumbent are

$$\hat{\sigma}_I = \begin{cases} (0,1) & \text{if } p < 1 \\ (y, 1-y) \text{ with } y \in [0,1] & \text{if } p = 1 \end{cases}$$

So the Nash equilibria are (E, A) and (S, σ^*) where $\sigma^* = (y, 1-y)$ with $y \geq \frac{1}{2}$.

5.7 In the subgame beginning at the right-hand decision node, player 2 will always choose R. The simultaneous decision subgame beginning at the second player's left-hand decision node has 3 Nash equilibria: (a) (C, C), (b) (D, D), and (c) (σ, σ) where $\sigma = (\frac{1}{2}, \frac{1}{2})$. These yield payoff pairs (a) (3,1), (b) (1,3), and (c) (1.5,0.5). So there are three subgame perfect Nash equilibria: (AC, CR), (AD, DR), and $(B\sigma, \sigma R)$.

5.8 Nash equilbria for the simultaneous choice subgame are (A, a), (B, b), and (σ^*, σ^*) with $\sigma^* = (\frac{1}{2}, \frac{1}{2})$. Because $\pi_i(\sigma^*, \sigma^*) = 2$ for $i = 1, 2$, the subgame perfect Nash equilibria are (Ra, A), (Rb, B) and $(L\sigma^*, \sigma^*)$.

5.9 The Nash equilibria for the simultaneous decision subgame are (A, a), (B, b), and (σ_1^*, σ_2^*) with $\sigma_1^* = (1/4, 3/4)$ and $\sigma_2^* = (1/6, 5/6)$. Payoffs for the mixed strategy equilbrium are $\pi_1(\sigma_1^*, \sigma_2^*) = 5/6$ and $\pi_2(\sigma_1^*, \sigma_2^*) = 3/4$. Therefore, the three subgame perfect Nash equilibria are $(L\sigma_1^*, r\sigma_2^*)$, (LB, rb), and $(RA, \ell a)$. The equilibrium $(RA, \ell a)$ is supported by the following forward induction argument. If player 1 plays R, then player 2 would reason that player 1 will play A in the simultaneous subgame because that is the only way they will get a payoff greater than 4. So player 2 knows that they should coordinate on the Nash equilibrium (A, a) in that subgame. Because player 1 "knows" that player 2 will reason in this way, they will indeed play R with the expectation of receiving 5 rather than the 4 that would be achieved for playing L.

5.10 Player 2 should choose L if their payoff for doing so exceeds the expected payoff for choosing R. That is, they should choose L if and only if $2 > 3\varepsilon + (1-\varepsilon)$, which reduces to $\varepsilon < \frac{1}{2}$. (i.e., $\bar{\varepsilon} = \frac{1}{2}$.)

Chapter 6

6.1 Payoffs are

$$\pi_i(q_1, q_2) = q_i \left[P_0 \left(1 - \frac{q_1 + q_2}{Q_0} \right) - c_i \right]$$

so the best responses are

$$\hat{q}_1 = \frac{Q_0}{2} \left(1 - \frac{q_2}{Q_0} - \frac{c_1}{P_0} \right) \quad \text{and} \quad \hat{q}_2 = \frac{Q_0}{2} \left(1 - \frac{q_1}{Q_0} - \frac{c_2}{P_0} \right).$$

Therefore, Nash equilibrium strategies are found by solving the simultaneous equations

$$q_1^* = \frac{Q_0}{2}\left(1 - \frac{q_2^*}{Q_0} - \frac{c_1}{P_0}\right)$$

$$q_2^* = \frac{Q_0}{2}\left(1 - \frac{q_1^*}{Q_0} - \frac{c_2}{P_0}\right)$$

which gives

$$q_1^* = \frac{Q_0}{3}\left(1 - \frac{2c_1 - c_2}{P_0}\right)$$

$$q_2^* = \frac{Q_0}{3}\left(1 - \frac{2c_2 - c_1}{P_0}\right).$$

For this to be a Nash equilibrium, we must also have $q_1^* > 0$ and $q_2^* > 0$, which implies that we must have $2c_1 - c_2 < P_0$ and $2c_2 - c_1 < P_0$. Suppose that $0 < c_1, c_2 < \frac{1}{2}P_0$, then these conditions are satisfied. So the pair of quantities q_1^* and q_2^* given above is indeed a Nash equilibrium. On the other hand, if $2c_2 > P_0 + c_1$ then $q_2^* < 0$ by the formula above, so it cannot be part of a Nash equilibrium. In this case, the Nash equilibrium is

$$q_1^* = \frac{Q_0}{2}\left(1 - \frac{c_1}{P_0}\right)$$

$$q_2^* = 0$$

because the q_1^* given above is the best response to $q_2 = 0$ from the equation for $\hat{q}_1$, and, given that Firm 1 is producing this quantity, the payoff for Firm 2 is maximised on the domain $0 \le q_2 < \infty$ at $q_2 = 0$.

6.2 Because we are looking for a symmetric Nash equilibrium, assume that all the firms except the first are producing a quantity q and Firm 1 is producing a (possibly different) quantity q_1. Then

$$\pi_1(q_1, q, q, \ldots, q) = q_1 \left[P_0\left(1 - \frac{q_1 + (n-1)q}{Q_0}\right) - c\right]$$

The best response for Firm 1 is then

$$\hat{q}_1 = \frac{Q_0}{2}\left(1 - (n-1)\frac{q}{Q_0} - \frac{c}{P_0}\right).$$

So the symmetric Nash equilibrium quantity q^* is

$$q^* = \frac{Q_0}{2}\left(1 - (n-1)\frac{q^*}{Q_0} - \frac{c}{P_0}\right)$$

$$= \frac{Q_0}{n+1}\left(1 - \frac{c}{P_0}\right)$$

This gives a profit to each firm of

$$\pi_i(q^*, q^*, \ldots, q^*) = q^*\left[P_0\left(1 - \frac{nq^*}{Q_0}\right) - c\right]$$

$$= \frac{Q_0 P_0}{(n+1)^2}\left(1 - \frac{c}{P_0}\right)^2.$$

So $\lim_{n \to \infty} \pi_i = 0$.

6.3 The payoffs are
$$B_1(e_1, e_2) = B_0(e_1 + ke_2)^2 + c(e_0 - e_1)$$
and
$$B_2(e_1, e_2) = B_0(e_2 + ke_1)^2 + c(e_0 - e_2).$$
Each country wants to minimise its B_i, so the best response for Country 1 to some fixed level of pollution e_2 is found by (check the second derivative)
$$\frac{\partial B_1}{\partial e_1}(\hat{e}_1, e_2) = 0$$
which gives
$$\hat{e}_1 = \frac{c}{2B_0} - ke_2.$$
By symmetry
$$\hat{e}_2 = \frac{c}{2B_0} - ke_1.$$
A Nash equilibrium is a pair of strategies e_1^* and e_2^* that are best responses to each other. Using symmetry, we must have $e_1^* = e_2^* = e^*$ where
$$e^* = \frac{c}{2B_0} - ke^*$$
$$= \frac{c}{2B_0(1+k)}$$
At equilibrium, the total amount of pollution in each country is
$$E^* = \frac{c}{2B_0}$$
which is independent of the amount of pollution coming from the adjacent country (i.e., the more one country affects the other, the larger is each country's abatement) and the total amount of pollution is high when costs of cleaning up are high.

6.4 The aggregate quantity produced in the Stackelberg model is
$$Q_S^* = q_1^* + q_2^* = \frac{3}{4}Q_0\left(1 - \frac{c}{P_0}\right)$$
which is larger than the aggregate quantity produced in the Cournot model
$$Q_C^* = 2q_C^* = \frac{2}{3}Q_0\left(1 - \frac{c}{P_0}\right)$$
which implies that the market price of a single item is smaller in the Stackelberg model than in the Cournot model.

6.5 In the event that the Entrant does diversify, the incumbent will make a greater profit if it reveals its production levels (i.e., in a Stackelberg duopoly), so the Entrant will be a market follower and will make a profit
$$\pi_2^* = \frac{P_0 Q_0}{16}\left(1 - \frac{c}{P_0}\right)^2.$$
The Entrant will diversify if potential profits exceed the cost of entering the market:
$$\frac{P_0 Q_0}{16}\left(1 - \frac{c}{P_0}\right)^2 > C_E.$$

6.6 The result can be found by explicit calculation of the expected payoff

$$\pi_1(\sigma_1^*, \sigma_2^*) = \int_0^\infty p(x) \left[\int_0^x (v-y)q(y)dy - x \int_x^\infty q(y)dy \right] dx .$$

or by observing that $\pi_1(0, \sigma_2^*) = 0$ together with the condition

$$\frac{\partial}{\partial x}\pi_1(x, \sigma_2^*) = 0$$

implies that $\pi_1(x, \sigma_2^*) = 0 \ \forall x$. Hence

$$\pi_1(\sigma_1, \sigma_2^*) = 0$$

for all σ_1 including $\sigma_1 = \sigma_1^*$.

6.7 Because the analysis in the text was done in terms of costs, we still have

$$p(x) = \frac{1}{v} \exp\left(-\frac{x}{v}\right) .$$

But

$$\begin{aligned}
p(t) &= p(x)\frac{dx}{dt} \\
&= \frac{2kt}{v} \exp\left(-\frac{kt^2}{v}\right)
\end{aligned}$$

which is not exponential and which, in turn, leads to a non-exponential distribution of contest durations.

Chapter 7

7.1 Ignoring a common factor of

$$Q_0 P_0 \left(1 - \frac{c}{P_0}\right)^2$$

the payoff table is

		Firm 2	
		M	C
Firm 1	M	$\frac{1}{8}, \frac{1}{8}$	$\frac{5}{48}, \frac{5}{36}$
	C	$\frac{5}{36}, \frac{5}{48}$	$\frac{1}{9}, \frac{1}{9}$

Setting

$$r = \frac{1}{8} \quad t = \frac{5}{36} \quad s = \frac{5}{48} \quad p = \frac{1}{9}$$

we see this game has the structure of a Prisoners' Dilemma because $t > r > p > s$.

7.2 Payoff table is

		P_2			
		CC	CD	DC	DD
	CC	6,6	3,8	3,8	0,10
P_1	CD	8,3	4,4	5,5	1,6
	DC	8,3	5,5	4,4	1,6
	DD	10,0	6,1	6,1	2,2

All strategies are dominated except DD for both players, so (DD, DD) is the *unique* Nash equilibrium (not just in pure strategies).

7.3 Comparison of $\pi_1(s_T, s_T)$, $\pi_1(s_C, s_T)$, and $\pi_1(s_D, s_T)$ leads to $\delta \geq \frac{1}{2}$ as in Example 7.4. But we must also compare $\pi_1(s_T, s_T)$ and $\pi_1(s_A, s_T)$. Because

$$\pi_1(s_A, s_T) = 5 + 0 + 5\delta^2 + 0 + 5\delta^4 + \ldots$$
$$= \frac{5}{1 - \delta^2}$$

we require $\delta \geq \frac{2}{3}$.

7.4 Ignoring the (irrelevant) common factor of $(1-\delta)^{-1}$ the payoff table is

		Player 2		
		s_D	s_C	s_G
	s_D	1,1	5,0	$5-4\delta,\delta$
Player 1	s_C	0,5	3,3	3,3
	s_G	$\delta, 5-4\delta$	3,3	3,3

The pair (s_D, s_D) is always a Nash equilibrium because $\delta < 1$. The pair (s_G, s_G) is a Nash equilibrium if $3 \geq 5 - 4\delta$, which reduces to $\delta \geq \frac{1}{2}$. For $\delta < \frac{1}{2}$, only s_D is undominated, so $[s_D, s_D]$ is the unique Nash equilibrium in this case. For $\delta \geq \frac{1}{2}$, the game is not generic.
Let $\sigma_1 = (p, q, 1-p-q)$ and $\sigma_2 = (r, s, 1-r-s)$. Then

$$\pi_1(\sigma_1, \sigma_2) = 3 - (3-\delta)r + p[2 - r - 4\delta + 3r\delta + 4s\delta] - q[r\delta]$$

So the best responses are (with $\hat{p}, \hat{q} \in [0, 1]$ and $\hat{p} + \hat{q} = 1$)

$$\hat{\sigma}_1 = \begin{cases} s_D & \text{if } r > 0 \text{ and } 2 - 4\delta + (3\delta - 1)r + 4\delta s > 0 \\ (\hat{p}, 0, 1-\hat{p}) & \text{if } r > 0 \text{ and } 2 - 4\delta + (3\delta - 1)r + 4\delta s = 0 \\ s_G & \text{if } r > 0 \text{ and } 2 - 4\delta + (3\delta - 1)r + 4\delta s < 0 \\ s_D & \text{if } r = 0 \text{ and } s > 1 - (2\delta)^{-1} \\ (\hat{p}, \hat{q}, 1-\hat{p}-\hat{q}) & \text{if } r = 0 \text{ and } s = 1 - (2\delta)^{-1} \\ (0, \hat{q}, 1-\hat{q}) & \text{if } r = 0 \text{ and } s < 1 - (2\delta)^{-1} \end{cases}$$

$\hat{\sigma}_2$ is similar with $p \leftrightarrow r$ and $q \leftrightarrow s$. Hence the Nash equilibria are
a) Always defect: (s_D, s_D).

b) A specific mixture of defection (s_D) and conditional cooperation (s_G): $((x, 0, 1 - x), (x, 0, 1 - x))$ with $x = (4\delta - 2)/(3\delta - 1)$.
c) A set of mixtures of conditional (s_G) and unconditional cooperation (s_C): $((0, q, 1 - q), (0, s, 1 - s))$ with $q, s \leq 1 - (2\delta)^{-1}$.

7.5 Changing from s_C to s_B will not increase the payoff for either player. Only changing to s_A has the potential to do this. The critical value of δ for player 1 is given by the inequality

$$\pi_1(s_C, s_C) \geq \pi_1(s_A, s_C)$$
$$\iff \frac{2}{1 - \delta} \geq 3 + \frac{\delta}{1 - \delta}$$
$$\iff \delta \geq \frac{1}{2}.$$

The critical value of δ for player 2 is given by the inequality

$$\pi_2(s_C, s_C) \geq \pi_2(s_C, s_A)$$
$$\iff \frac{3}{1 - \delta} \geq 5 + \frac{2\delta}{1 - \delta}$$
$$\iff \delta \geq \frac{2}{3}.$$

Because the actual discount factor is common to both players, we must have

$$\delta \geq \max\left(\frac{1}{2}, \frac{2}{3}\right) = \frac{2}{3}$$

for (s_C, s_C) to be a Nash equilibrium.

7.6 Let s_T denote the Tit-for-Tat strategy. Now consider a stage t when player 2 defects but player 1 does not. In stage $t + 1$, the Nash equilibrium (s_T, s_T) specifies that player 1 should defect and player 2 should cooperate. These behaviours are then reversed for stage $t+2$, and so on. So in the subgame starting at stage $t + 1$, player 2 uses the strategy s_T and player 1 uses the cautious version of Tit-for-Tat (s_A), which begins by defecting rather than cooperating (see Exercise 7.3). Because $\delta > 2/3$, (s_A, s_T) is not a Nash equilibrium for the subgame starting at stage $t + 1$, as

$$\pi_1(s_A, s_T) = 5 + 5\delta^2 + 5\delta^4 + \ldots$$
$$= \frac{5}{1 - \delta^2}$$
$$< \frac{3}{1 - \delta}$$
$$= \pi_1(s_T, s_T)$$

for $\delta > \frac{2}{3}$.

7.7 Because the strategy s_P only depends on the behaviour of the players in the previous stage, we consider the possible behaviours at state $t - 1$ and examine what happens if player 1 deviates from s_P at stage t. (The game is symmetric so we don't need to consider player 2 separately.)
Consider the case when one of the players has used D at stage $t - 1$ and the other has used C (it does not matter which). Then s_P specifies using D in

stage t (for both players). The total future payoff to player 1 (including that from stage t) is then

$$\pi_1(s_P, s_P) = 1 + \frac{4\delta}{1-\delta}.$$

Suppose that, instead of using D, player 1 uses C in stage t and then reverts to s_P for stages $t+1$ onwards. Let us denote this strategy by s'. The total future payoff to player 1 is then

$$\pi_1(s', s_P) = 0 + \delta + \frac{4\delta^2}{1-\delta}.$$

Player 1 does not benefit from the switch if $\pi_1(s_P, s_P) \geq \pi_1(s', s_P)$, which is true for all values of δ.

Consider the case when both players have used D at stage $t-1$ or both have used C (it does not matter which). Then s_P specifies using C in stage t (for both players). The total future payoff to player 1 (including that from stage t) is then

$$\pi_1(s_P, s_P) = \frac{4}{1-\delta}.$$

Suppose that, instead of using C, player 1 uses D in stage t and then reverts to s_P for stages $t+1$ onwards. Let us denote this strategy by s''. The total future payoff to player 1 is then

$$\pi_1(s'', s_P) = 5 + \delta + \frac{4\delta^2}{1-\delta}.$$

Player 1 does not benefit from the switch if $\pi_1(s_P, s_P) \geq \pi_1(s'', s_P)$, which is true if $4 + 4\delta \geq 5 + \delta$. Consequently, (s_P, s_P) is a subgame perfect Nash equilibrium if $\delta \geq \frac{1}{3}$.

7.8 A suitable stochastic game is shown below.

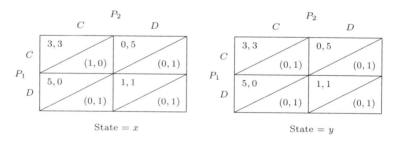

The game starts in state x and the Markov equivalent to σ_g is the pair $\{a(x) = C, a(y) = D\}$ (which we will write as CD).

In state y, the effective game is just the Prisoners' Dilemma itself so the equilibrium is for both players to use D. So we have

$$\pi_i^*(y) = \frac{1}{1-\delta}.$$

In state x, the effective game is

… Solutions

		P_2	
		C	D
P_1	C	$3+\delta\pi_1^*(x), 3+\delta\pi_2^*(x)$	$\delta\pi_1^*(y), 5+\delta\pi_2^*(y)$
	D	$5+\delta\pi_1^*(y), \delta\pi_2^*(y)$	$1+\delta\pi_1^*(y), 1+\delta\pi_2^*(y)$

or

		P_2	
		C	D
P_1	C	$3+\delta\pi_1^*(x), 3+\delta\pi_2^*(x)$	$\frac{\delta}{1-\delta}, 5+\frac{\delta}{1-\delta}$
	D	$5+\frac{\delta}{1-\delta}, \frac{\delta}{1-\delta}$	$\frac{1}{1-\delta}, \frac{1}{1-\delta}$

Clearly (D,C) and (C,D) are never an equilibrium of this effective game, and (D,D) is an equilibrium for all values of δ. The pair of actions (C,C) is an equilibrium if (for $i=1,2$)

$$3+\delta\pi_i(x) \geq 5+\frac{\delta}{1-\delta}$$

which means that we would have

$$\pi_i(x) \geq \frac{2-\delta}{\delta(1-\delta)}.$$

Suppose that both players choose cooperation in state x. Then

$$\pi_i(x) = 3+\delta\pi_i(x)$$
$$\Longrightarrow \pi_i(x) = \frac{3}{1-\delta}.$$

Now

$$\frac{3}{1-\delta} \geq \frac{2-\delta}{\delta(1-\delta)} \iff \delta \geq \frac{1}{2}$$

so (CD, CD) is a Markov-strategy Nash equilibrium if $\delta \geq \frac{1}{2}$.

Chapter 8

8.1 (a) There are two obvious ways: (i) 100% of the population uses the mixed strategy $\left(\frac{5}{12}, \frac{7}{12}\right)$; (ii) $\frac{5}{12}$ of the population use the pure strategy $(1,0)$ and $\frac{7}{12}$ use the pure strategy $(0,1)$. There are many possible, less obvious alternatives. For example, $\frac{1}{3}$ of the population uses the mixed strategy $\left(\frac{3}{4}, \frac{1}{4}\right)$ and $\frac{2}{3}$ use the mixed strategy $\left(\frac{1}{4}, \frac{3}{4}\right)$.
(b) $\mathbf{x} = \frac{4}{10}\left(\frac{1}{2}, 0, \frac{1}{2}\right) + \frac{6}{10}\left(\frac{1}{4}, \frac{3}{4}, 0\right) = \left(\frac{7}{20}, \frac{9}{20}, \frac{4}{20}\right).$

8.2 Candidate ESSs are
σ_W : Everyone uses W, then $x=1$ and $\pi(W,1) > \pi(L,1)$.
σ_L : Everyone uses L, then $x=0$ and $\pi(L,0) > \pi(W,0)$.

σ_m : A mixed strategy in which W is used $\frac{3}{4}$ of the time, then $x = \frac{3}{4}$ and $\pi(W, \frac{3}{4}) = \pi(L, \frac{3}{4})$.
Now $\mathbf{x}_\varepsilon = (p^* + \varepsilon(p - p^*), 1 - p^* - \varepsilon(p - p^*))$. So

$$\begin{aligned} \delta\pi &\equiv \pi(\sigma^*, \mathbf{x}_\varepsilon) - \pi(\sigma, \mathbf{x}_\varepsilon) \\ &= p^*\pi(W, \mathbf{x}_\varepsilon) + (1-p^*)\pi(L, \mathbf{x}_\varepsilon) - p\pi(W, \mathbf{x}_\varepsilon) - (1-p)\pi(L, \mathbf{x}_\varepsilon) \\ &= (p^* - p)\left(\pi(W, \mathbf{x}_\varepsilon) - \pi(L, \mathbf{x}_\varepsilon)\right) \\ &= (p^* - p)(4p^* - 3 - 4\varepsilon(p^* - p)) \end{aligned}$$

Taking each of the candidate ESSs in turn, we have
σ_W : $p^* = 1$, so $\delta\pi = (1-p)(1 - 4\varepsilon(1-p)) > 0 \quad \forall p \neq 1$ and for $\varepsilon < \bar{\varepsilon} = \frac{1}{4}$. So σ_W is an ESS.
σ_L : $p^* = 0$, so $\delta\pi = p(3 - 4\varepsilon p) > 0 \quad \forall p \neq 0$ and for $\varepsilon < \bar{\varepsilon} = \frac{3}{4}$. So σ_L is an ESS.
σ_m : $p^* = \frac{3}{4}$, so $\delta\pi = -4\varepsilon(\frac{3}{4} - p)^2 < 0 \quad \forall p \neq \frac{3}{4}$ and $\forall \varepsilon > 0$. So σ_m is not an ESS.

8.3 (a) Each female child gets 1 mating with n offspring per mating and each male child gets $(1-\mu)/\mu$ matings with n offspring per mating. So the expected number of grandchildren for a female using $\sigma = (p, 1-p)$ is

$$n^2 \left[p\left(0.8\frac{1-\mu}{\mu} + 0.2\right) + (1-p)\left(0.2\frac{1-\mu}{\mu} + 0.8\right)\right]$$

which simplifies to

$$\frac{n^2}{5}\left[\frac{3\mu+1}{\mu} + 3p\left(\frac{1-2\mu}{\mu}\right)\right]$$

(b) Because $\mu = \frac{1}{2}$ cannot be produced by either pure strategy, it must be produced by a mixed strategy. If this mixed strategy is an ESS, then its payoff must be independent of p (or, equivalently, the payoffs to the two pure strategies must be the same). From the expression given above, this requires $\mu = \frac{1}{2}$.

(c) The sex ratio produced by a strategy $\sigma = (p, 1-p)$ is

$$\mu = 0.8p + 0.2(1-p) = \frac{1}{5}(1 + 3p).$$

So to produce a sex ratio of $\mu = \frac{1}{2}$ we must have $p^* = \frac{1}{2}$ and $\sigma^* = (\frac{1}{2}, \frac{1}{2})$. To prove that σ^* is an ESS, we need to check that

$$\pi(\sigma^*, \mathbf{x}_\varepsilon) > \pi(\sigma, \mathbf{x}_\varepsilon) \quad \forall \sigma \neq \sigma^*$$

where $\mathbf{x}_\varepsilon = (1-\varepsilon)\sigma^* + \varepsilon\sigma$. Let $\sigma = (p, 1-p)$ with $p \neq p^*$. Then $\mathbf{x}_\varepsilon = (p_\varepsilon, 1-p_\varepsilon)$ with $p_\varepsilon = (1-\varepsilon)p^* + \varepsilon p$, which leads to a proportion of males in the population

$$\mu_\varepsilon = \frac{1}{5}(1 + 3p_\varepsilon)$$

Now,

$$\pi(\sigma^*, \mathbf{x}_\varepsilon) > \pi(\sigma, \mathbf{x}_\varepsilon)$$
$$\iff (p^* - p)\left(\frac{1-2\mu_\varepsilon}{\mu_\varepsilon}\right) > 0$$

However, if $p > 0.5$ then $p_\varepsilon > 0.5$ and, hence, $\mu_\varepsilon > 0.5$. Conversely, if $p < 0.5$ then $p_\varepsilon < 0.5$ and, hence, $\mu_\varepsilon < 0.5$. So the inequality is satisfied for any $p \neq p^*$, and hence σ^* is an ESS.

8.4 Let p be the probability of playing H, then $\sigma = (p, 1-p)$, $\sigma^* = (1,0)$ and $\mathbf{x}_\varepsilon = (1 - \varepsilon + \varepsilon p, \varepsilon - \varepsilon p)$.

$$\pi(\sigma^*, \nu_\varepsilon) = (1 - \varepsilon + \varepsilon p)\left(\frac{v-c}{2}\right) + \varepsilon(1-p)v$$

$$\pi(\sigma, \nu_\varepsilon) = p(1 - \varepsilon + \varepsilon p)\left(\frac{v-c}{2}\right) + p\varepsilon(1-p)v + \varepsilon(1-p)^2\frac{v}{2}$$

So

$$\pi(\sigma^*, \mathbf{x}_\varepsilon) - \pi(\sigma, \mathbf{x}_\varepsilon) = (1-p)\left[\frac{v-c}{2} + \varepsilon(1-p)\frac{c}{2}\right]$$
$$> 0 \quad \forall p \neq 1$$

because $v \geq c$.

8.5 Let p be the probability of cooperating (i.e., playing C), then $\sigma = (p, 1-p)$, $\sigma^* = (0,1)$, and $\mathbf{x}_\varepsilon = (\varepsilon p, 1 - \varepsilon p)$. Then

$$\pi(\sigma^*, \mathbf{x}_\varepsilon) = 1 + 4\varepsilon p$$
$$\pi(\sigma, \mathbf{x}_\varepsilon) = (1-p) + \varepsilon p(4-p)$$

So

$$\pi(\sigma^*, \mathbf{x}_\varepsilon) - \pi(\sigma, \mathbf{x}_\varepsilon) = p(1 + \varepsilon p)$$
$$> 0 \quad \forall p \neq 0$$

8.6 (a) The three pure strategies $R \equiv (1,0,0)$, $G \equiv (0,1,0)$, and $B \equiv (0,0,1)$ are all ESSs. The mixed strategy Nash equilibrium with $\sigma^* = \left(\frac{1}{3}, \frac{1}{3}, \frac{1}{3}\right)$ is not an ESS.
(b) $G \equiv (1,0)$ is the only ESS, because $\pi(G,G) > \pi(H,G)$.
(c) $A \equiv (1,0)$ and $B \equiv (0,1)$ are both ESSs. The mixed-strategy Nash equilibrium $\sigma^* = (p^*, 1-p^*)$ with $p^* = \frac{2}{5}$ is not an ESS because

$$\pi(\sigma^*, \sigma) - \pi(\sigma, \sigma) = -5(p-p^*)^2 .$$

(d) The strategy $\sigma^* = (2/3, 1/3)$ is the unique ESS because

$$\pi(\sigma^*, \sigma) - \pi(\sigma, \sigma) = 4 - 12p - 9p^2$$

which is positive for all $p \neq \frac{2}{3}$.

8.7 (a) The payoff table is

	T_1	T_2
T_1	2,2	1,1
T_2	1,1	2,2

By inspection, (T_1, T_1) and (T_2, T_2) are symmetric Nash equilibria. Find mixed strategy Nash equilibria using the equality of payoffs theorem.

$$\pi_1(A, \sigma_2^*) = \pi_1(B, \sigma_2^*)$$
$$\Longrightarrow 2q^* + (1-q^*) = q^* + 2(1-q^*)$$
$$\Longrightarrow q^* = \frac{1}{2}.$$

By symmetry $p^* = \frac{1}{2}$, so the mixed strategy Nash equilibrium is $\left(\left(\frac{1}{2}, \frac{1}{2}\right), \left(\frac{1}{2}, \frac{1}{2}\right)\right)$.
(b) Both T_1 and T_2 are ESSs for the following reasons. Let $\sigma = (p, 1-p)$, so T_1 corresponds to $p=1$ and T_2 corresponds to $p=0$.

- $\pi(T_1, T_1) = 2$ and $\pi(\sigma, T_1) = 1 + p$. Hence $\pi(T_1, T_1) > \pi(\sigma, T_1)$ $\forall p \neq 1$.
- $\pi(T_2, T_2) = 2$ and $\pi(\sigma, T_2) = 2 - p$. Hence $\pi(T_2, T_2) > \pi(\sigma, T_2)$ $\forall p \neq 0$.

The mixed strategy $\sigma^* = (\frac{1}{2}, \frac{1}{2})$ is not an ESS because (for example)

$$\pi(\sigma^*, \sigma^*) = \pi(T_1, \sigma^*) = \frac{3}{2} \quad \forall \sigma \neq \sigma^*$$

but $\pi(T_1, T_1) = 2$.

8.8 Let w, x, y, and z be the probabilities of playing HH, HD, DH, and DD, respectively, at a mixed strategy Nash equilibrium $[\sigma^*, \sigma^*]$ with $\sigma^* = (w, x, y, z)$. Then

$$\begin{aligned}
\pi(HH, \sigma^*) &= -2w + x + y + z \\
\pi(HD, \sigma^*) &= -w + 2x + 3z \\
\pi(DH, \sigma^*) &= -w + 2y + 3z \\
\pi(DD, \sigma^*) &= x + y + 2z
\end{aligned}$$

Equating these payoffs in all possible combinations and using the constraint $w + x + y + z = 1$ gives $w = z$, $x = y$ and $w + x = \frac{1}{2}$. Hence

$$\pi(HH, \sigma^*) = \pi(HD, \sigma^*) = \pi(DH, \sigma^*) = \pi(DD, \sigma^*) = \pi(\sigma^*, \sigma^*) = 1.$$

Now $\pi(\sigma^*, HD) = w + 2x + z = 1$ but $\pi(HD, HD) = 2$ so σ^* is not an ESS.

8.9 (a) This game has no ESSs, because the payoff is the same for all possible strategies.
(b) There are no pure-strategy ESSs. The symmetric mixed-strategy Nash equilibrium has $\sigma^* = (\frac{1}{2}, \frac{1}{2})$. Because $\pi(\sigma^*, \sigma) = \frac{1}{2} + p$ and $\pi(\sigma, \sigma) = 3p - 2p^2$ (where p is the probability of playing E), we have $\pi(\sigma^*, \sigma) > \pi(\sigma, \sigma)$ $\forall p \neq \frac{1}{2}$.
(c) There are no pure-strategy ESSs. The mixed-strategy Nash equilibrium with $\sigma^* = (\frac{1}{3}, \frac{1}{2}, \frac{1}{6})$ is also not an ESS because $\pi(\sigma^*, \sigma) = \pi(\sigma, \sigma) = 0$ $\forall \sigma$.

Chapter 9

9.1 Because

$$\begin{aligned}
\sum_{i=1}^{k} \dot{x}_i &= (\pi(s_i, \mathbf{x}) - \bar{\pi}(\mathbf{x}))x_i \\
&= \sum_{i=1}^{k} (\pi(s_i, \mathbf{x})x_i - \bar{\pi}(\mathbf{x})) \sum_{i=1}^{k} x_i \\
&= \bar{\pi}(\mathbf{x})) - \bar{\pi}(\mathbf{x})) \\
&= 0
\end{aligned}$$

the result follows.

9.2 Under the affine transformation $\bar{\pi}(\mathbf{x}) \to \lambda\bar{\pi}(\mathbf{x}) - \mu$, so Equation 9.1 becomes
$$\frac{dx_i}{dt} = \lambda((\pi(s_i, \mathbf{x}) - \bar{\pi}(\mathbf{x}))x_i\,.$$
Introducing an adjusted time parameter $\tau = \lambda t$, we can write this as
$$\frac{dx_i}{d\tau} = ((\pi(s_i, \mathbf{x}) - \bar{\pi}(\mathbf{x}))x_i$$
which is exactly the same form as the original equation.

9.3 Because
$$\begin{aligned}\pi(A, \mathbf{x}) &= (a-b)x_1 + 2ax_2 \\ \pi(B, \mathbf{x}) &= ax_2\end{aligned}$$
the average payoff is $\bar{\pi}(\mathbf{x}) = (a-b)x_1^2 + 2ax_1x_2 + ax_2^2$ and the replicator dynamics is
$$\begin{aligned}\dot{x}_1 &= x_1((a-b)x_1 + 2ax_2 - \bar{\pi}(\mathbf{x})) \\ \dot{x}_2 &= x_2(ax_2 - \bar{\pi}(\mathbf{x}))\,.\end{aligned}$$
Clearly the populations $(x_1 = 1, x_2 = 0)$ and $(x_1 = 0, x_2 = 1)$ are fixed points. At the polymorphic fixed point, we must have
$$(a-b)x_1 + 2ax_2 - \bar{\pi}(\mathbf{x}) = 0 = ax_2 - \bar{\pi}(\mathbf{x})$$
which gives $(a-b)x_1 = -ax_2$. Substituting this into the equation $ax_2 - \bar{\pi}(\mathbf{x}) = 0$ gives $x_1 = \frac{a}{b}$.

9.4 Let x be the proportion of H-players, then
$$\dot{x} = \frac{c}{2}x(1-x)\left(\frac{v}{c} - x\right)$$
with fixed points $x^* = 0$, $x^* = 1$, and $x^* = v/c$. If $x < v/c$, then $\dot{x} > 0$ and if $x > v/c$, then $\dot{x} < 0$. So $x \to v/c$ for any initial population that is not at a fixed point.

9.5 The replicator dynamics equation for the proportion of T_1-players is
$$\begin{aligned}\dot{x} &= x(1-x)\left(\pi(T_1, \mathbf{x}) - \pi(T_2, \mathbf{x})\right) \\ &= x(1-x)(2x-1)\,.\end{aligned}$$
If $x > \frac{1}{2}$, then $x \to 1$ and if $x < \frac{1}{2}$, then $x \to 0$. From Exercise 8.7 the ESSs are T_1 and T_2, which correspond to these evolutionary end points.

9.6 When $a > 0$, both A and B are ESSs. For $a < 0$, the game has a unique ESS, $\sigma^* = (1/2, 1/2)$. The replicator dynamics equation is
$$\dot{x} = ax(1-x)(2x-1)$$
with fixed points $x^* = 0$, $x^* = 1$ and $x^* = \frac{1}{2}$.
First, consider a population near to $x^* = 0$. Let $x = x^* + \varepsilon = \varepsilon$. Then we have
$$\begin{aligned}\dot{\varepsilon} &= a\varepsilon(1-\varepsilon)(2\varepsilon - 1) \\ &\approx -a\varepsilon\,.\end{aligned}$$

So the fixed point $x^* = 0$ is asymptotically stable if $a > 0$ and unstable if $a < 0$.

Now consider a population near to $x^* = 1$. Let $x = x^* - \varepsilon = 1 - \varepsilon$. Then we have

$$\begin{aligned}\dot{\varepsilon} &= -a(1-\varepsilon)(\varepsilon)(2(1-\varepsilon)-1) \\ &\approx -a\varepsilon\end{aligned}$$

So $x^* = 1$ is asymptotically stable if $a > 0$ and unstable if $a < 0$.

Finally, consider a population near to $x^* = \frac{1}{2}$. Let $x = x^* + \varepsilon = \frac{1}{2} + \varepsilon$. Then we have

$$\begin{aligned}\dot{\varepsilon} &= a(\frac{1}{2}+\varepsilon)(1-\frac{1}{2}-\varepsilon)(1+2\varepsilon-1) \\ &\approx \frac{1}{2}a\varepsilon\ .\end{aligned}$$

So $x_3^* = \frac{1}{2}$ is asymptotically stable if $a < 0$ and unstable if $a > 0$.

Overall a fixed point is asymptotically stable if and only if the corresponding strategy is an ESS.

9.7 The fixed points are $(1,0,0)$, $(0,1,0)$, and $(0,0,1)$ in both cases.

9.8 The replicator dynamics equations are

$$\begin{aligned}\dot{x} &= x(1+2x-y-\bar{\pi}(x,y)) \\ \dot{y} &= y(1-x+2y-\bar{\pi}(x,y))\end{aligned}$$

with

$$\bar{\pi}(x,y) = 1 + 2x^2 + 2y^2 - 2xy\ .$$

The fixed points are $(0,0)$, $(0,1)$, $(1,0)$, and $(\frac{1}{2}, \frac{1}{2})$. The points $(0,1)$ and $(1,0)$ are stable nodes (eigenvalues -2 and -3 in both cases). The point $(\frac{1}{2}, \frac{1}{2})$ is a saddle point (eigenvalues $\frac{1}{2}$ and $-\frac{3}{2}$ with eigenvectors $x+y=1$ and $x=y$, respectively). The point $(0,0)$ is non-hyperbolic. On the invariant lines $x=0$, $y=0$, and $x=y$, the population moves away from $(0,0)$. So a qualitative picture of the replicator dynamics looks like the figure below.

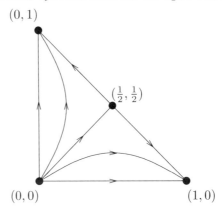

9.9 Let x be the proportion of A-players and let y be the proportion of B-players. Set $x = \frac{1}{3} + \xi$ and $y = \frac{1}{3} + \eta$ and linearise about the fixed point $x^* = y^* = \frac{1}{3}$ to get

$$\begin{pmatrix} \dot\xi \\ \dot\eta \end{pmatrix} = L \begin{pmatrix} \xi \\ \eta \end{pmatrix}$$

with

$$L = \frac{1}{9} \begin{pmatrix} 1 & 2 \\ -8 & -7 \end{pmatrix}$$

The eigenvalues of the matrix L are both $-\frac{1}{3}$. Hence the fixed point is asymptotically stable. Because $\sigma^* = (\frac{1}{3}, \frac{1}{3}, \frac{1}{3})$ is a mixed Nash equilibrium strategy, we have, for $\sigma = (0, \frac{1}{2}, \frac{1}{2})$,

$$\pi(\sigma, \sigma^*) = \pi(\sigma^*, \sigma^*) = \frac{2}{3}.$$

But

$$\pi(\sigma^*, \sigma) = \frac{5}{6}$$
$$< 1$$
$$= \pi(\sigma, \sigma)$$

so σ^* is not an ESS.

9.10 The payoff table is ($c > 0$)

	R	S	P
R	$-c,-c$	1,-1	-1,1
S	-1,1	$-c,-c$	1,-1
P	1,-1	-1,1	$-c,-c$

Let x, y, and z be the proportions of R-, S-, and P-players. Then the replicator dynamics system is

$$\dot x = x(-cx + y - z - \bar\pi(\mathbf{x}))$$
$$\dot y = y(-x - cy + z - \bar\pi(\mathbf{x}))$$
$$\dot z = z(x - y - cz - \bar\pi(\mathbf{x}))$$

with $\bar\pi(\mathbf{x}) = -c(x^2 + y^2 + z^2)$. It is easy to check that the point $x = y = z = \frac{1}{3}$ is a fixed point. Let V be the relative entropy function, then

$$\frac{dV}{dt} = -[\pi(\sigma^*, \mathbf{x}) - \bar\pi(\mathbf{x})]$$
$$= \frac{c}{3} - c(x^2 + y^2 + z^2)$$
$$< 0 \quad \text{for } \mathbf{x} \neq \left(\frac{1}{3}, \frac{1}{3}, \frac{1}{3}\right).$$

Further Reading

Part I

A detailed, technical exposition of utility theory is given by Myerson (1991); Allingham (2002) provides a more conceptual account. The philosophical background to rational behaviour is explored by Hargreaves Heap & Varoufakis (1995). Grafen (1991) gives a biological introduction to modelling animal behaviour, and the mathematical foundations for the concept of fitness are discussed by Houston & McNamara (1999). Markov decision processes are covered in detail by Ross (1995) and Puterman (1994). Biological applications of such processes are discussed by Mangel & Clark (1988).

Part II

Myerson (1991) and Fudenberg & Tirole (1993) give theoretical introductions to game theory and include many ideas not covered in this book. Good sources of game-theoretic models include Gibbons (1992), Gintis (2000), and Romp (1997). Brams (1983) provides a highly unusual, but thought-provoking, application of game theory. Game-theoretic models with continuous strategy sets are discussed by Gabszewicz (1999) and Martin (1993). The various Nash equilibrium refinements are discussed in detail by van Damme (1991). Stochastic games are covered by Filar & Vrieze (1997).

Part III

The classic text on evolutionary game theory was written by Maynard Smith (1982). More recent, biologically-oriented texts include those by Dugatkin & Reeve (1998) and Houston & McNamara (1999). The evolution of the Social Contract is discussed by Skyrms (1996). Replicator dynamics and other forms of evolutionary dynamics are covered by Weibull (1995), Vega-Redondo (1996), and Hofbauer & Sigmund (1998). Young (1998) emphasises stochastic evolutionary dynamics.

Bibliography

Allingham, M. (2002), *Choice Theory: A Very Short Introduction*, Oxford University Press, Oxford.

Brams, S. J. (1983), *Superior Beings: If They Exist, How Would We Know?*, Springer-Verlag, New York.

Dugatkin, L. A. & Reeve, H. K. (1998), *Game Theory and Animal Behavior*, Oxford University Press, Oxford.

Filar, J. & Vrieze, K. (1997), *Competitive Markov Decision Processes*, Springer-Verlag, Berlin.

Fudenberg, D. & Tirole, J. (1993), *Game Theory*, The MIT Press, Cambridge.

Gabszewicz, J. J. (1999), *Strategic Interaction and Markets*, Oxford University Press, Oxford.

Gibbons, R. (1992), *A Primer in Game Theory*, Harvester Wheatsheaf, Hemel Hempstead.

Gintis, H. (2000), *Game Theory Evolving*, Priceton University Press, Princeton.

Grafen, A. (1991), Modelling in behavioural ecology, *in* 'Behavioural Ecology: An Evolutionary Approach', 3 edn, Blackwell, Oxford, pp. 5–31.

Hargreaves Heap, S. P. & Varoufakis, Y. (1995), *Game Theory: A Critical Introduction*, Routledge, London.

Hofbauer, J. & Sigmund, K. (1998), *Evolutionary Games and Population Dynamics*, Cambridge University Press, Cambridge.

Houston, A. I. & McNamara, J. M. (1999), *Models of Adaptive Behaviour: An approach based on state*, Cambridge University Press, Cambridge.

Mangel, M. & Clark, C. W. (1988), *Dynamic Modeling in Behavioral Ecology*, Princeton University Press, Princeton.

Martin, S. (1993), *Advanced Industrial Economics*, Blackwell, Oxford.

Maynard Smith, J. (1982), *Evolution and the Theory of Games*, Cambridge University Press, Cambridge.

Myerson, R. B. (1991), *Game Theory: Analysis of Conflict*, Harvard University Press, Cambridge.

Puterman, M. L. (1994), *Markov Decision Processes: Discrete Stochastic Dynamic Programming*, Wiley, New York.

Romp, G. (1997), *Game Theory: Introduction and Applications*, Oxford University Press, Oxford.

Ross, S. M. (1995), *Introduction to Stochastic Dynamic Programming*, Academic Press, London.

Skyrms, B. (1996), *Evolution of the Social Contract*, Cambridge University Press, Cambridge.

van Damme, E. (1991), *Stability and Perfection of Nash Equilibria*, 2 edn, Springer-Verlag, Berlin.

Vega-Redondo, F. (1996), *Evolution, Games and Economic Behaviour*, Oxford University Press, Oxford.

Von Neuman, J. & Morgenstern, O. (1953), *Theory of Games and Economic Behavior*, 3 edn, Princeton University Press, Princeton.

Weibull, J. W. (1995), *Evolutionary Game Theory*, The MIT Press, Cambridge.

Young, H. P. (1998), *Individual Strategy and Social Structure: An Evolutionary Theory of Institutions*, Priceton University Press, Princeton.

Index

absorbing state   44, 133
action   5, 91
– optimal   6, 8, 9
affine transformation   6, 7, 15, 80, 81, 161, 167
argmax   4, 70
asymptotic stability   171–173, 181, 183, 197, 202

backward induction   34, 39, 40, 90, 92, 96, 99, 100, 112, 121
Battle of the Sexes   78
behaviour
– optimal   21
– randomising   21, 27, 38, 75
– support   22

cartel   110, 119
common knowledge of rationality   67
conventions   79
coordination game   78
correspondence   5

decision
– optimal   3, 6
decision node   23, 89
decision process   23, 31, 32
– deterministic   37
– state-dependent   37
– stochastic   38
decision tree   23, 24
Dinner Party   89
discounting   49, 123
dominated strategies

– elimination   66, 68
– iterated elimination   67
– strictly   66
– weakly   66
duopoly model
– Bertrand   108
– Cournot   108–111, 119, 120
– finite   120
– Stackelberg   111–113
dynamic programming   39, 41
– deterministic   41
– stochastic   44
dynamical systems
– linear   193–197
– non-linear   198–203

eigenvalues   177, 195
eigenvectors   177, 195
errors   104, 105
ESS   140, 144, 147, 151, 154, 165
– alternative stability concepts   163
– asymmetric game   158, 159
– definition   144
– existence   160–163
– finding   146, 149, 151, 152, 154, 156
– in dynamic games   163
– relation to Nash equilibrium   153–154
evolution   79
– in biology   139
– in economics   140
evolutionarily stable strategy   see ESS
evolutionary game theory   139
expected monetary value   11, 12
expected utility   15

fitness 19, 140, 145
fixed points 167, 169
– classification 177, 197
– hyperbolic 177, 201
– non-hyperbolic 178, 201
folk theorem 129–132
forward induction 102, 103

game
– dynamic 89
– effective 133, 134
– extensive form 89
– generic 82
– non-generic 82, 83
– normal form 64, 91
– static 61, 63
  strategic form 64, 91
– zero-sum 61, 84
game against the field 141
game tree 89

Hartman-Grobman theorem 177, 201
Hawk-Dove game 148–150, 155, 170
– asymmetric 158
history 38
– full 33
– null 33
– partial 33
horizon
– finite 37
– infinite 37

information set 93
invariant manifolds 176
Iterated Prisoners' Dilemma 120–125, 163

Lagrange multiplier 189, 190
Lagrangian 40, 189–191
linearisation 171, 176, 193, 198
lottery
– compound 14
– simple 13
Lyapounov function 181, 183, 202, 203

Markov decision process 39, 42
Markov decision processes
– finite 42–48
– infinite 48–57
Markov property 39
Matching Pennies 71
multiple equilibria 78, 79

Nash equilibrium 69, 70, 114, 154

– and dominated strategies 70, 71
– continuous strategy sets 107, 109, 115
– finding 71–76
– for n-player games 87
– for dynamic games 91
– for zero-sum games 85
– mass action interpretation 139, 180
– refinements 79, 101
– trembling hand 104
Nash's theorem 76
Natural Selection 17–19, 148

oddness theorem 82
one-stage deviation principle 127–129
optimality principle 32–34, 39
optimisation 3–5
– constrained 39, 189–192

pairwise contest 142, 154, 168
– associated game 153
– asymmetric 157–160
Pareto optimality 62
payoff 5
– uncertain 9
phenotypic gambit 18, 140
policy improvement 54–57
population games 139–163
population profile 140
post-entry population 143, 150–152
preference 12
Prisoners' Dilemma 61, 63, 119, 120, 130, 168

rôle asymmetry 158, 159
rationality 12, 14, 65
– under certainty 13
– under uncertainty 14
relative entropy 182, 184
repeated games 119, 120
– average payoffs 130
– feasible payoffs 129, 130
– individually rational payoffs 130
replicator dynamics 165
– and ESSs 169–173, 182, 183
– and Nash equilibria 169, 179, 181
– reduced state 174
Rock-Scissors-Paper game 160, 183, 185

sex ratio 144–147
Social Contract 63
social efficiency 63
stage game 120

stochastic games 132–135
strategy 23, 24, 91
- behavioural 27, 28, 47, 96
- best response 70
- conditional 120, 122, 158
- dominated *see* dominated strategies
- equivalent representations 28–30, 96
- Markov 46, 132, 133
- mixed 27, 28, 64, 96
- optimal 32, 40, 45
- outcome of 25
- pure 24–26
- state-dependent 120
- stationary 122
- support 28, 74
subgame 99
subgame perfection 99–102, 112, 121, 125–129, 131

Tit-for-Tat 125, 126, 163

utility 13
- attitude to risk 16

War of Attrition 114–117, 156

Printed in the United States of America.